#66973
11^{50}

New Uses of Systems Theory in Archaeology

AF478709

New Uses
of
Systems Theory
in
Archaeology

Edited by

E. Gary Stickel

Ballena Press
381 First Street
Los Altos, California 94022

Library of Congress Cataloging in Publication Data

Main entry under title:

New uses of systems theory in archaeology.

   (Ballena Press anthropological papers ; no. 24)
   Bibliography: p.
   1. Archaeology- -Philosophy- -Addresses, essays,
lectures. 2. Archaeology- -Methodology- -Addresses,
essays, lectures. 3. System theory- -Addresses, essays,
lectures. I. Stickel, E. Gary. II. Series.
CC72.N48 1982   930.1'01   82-13811
ISBN 0-87919-096-5

Copyright © 1982 by Ballena Press

All rights reserved. No part of this book may be reproduced in any form
or by any means without prior written permission of the publisher, ex-
cepting brief quotes used in connection with reviews written specifically
for inclusion in a magazine or newspaper.

Printed in the United States of America.
1st Printing.

CONTENTS

ILLUSTRATIONS

TABLES

<u>FIGURES</u>

Cover design by Gary Stickel and Claudine Scoville, based on a sculpture
mobile by Alexander Calder and on a central African mask; it symbolizes
the mutual interaction and dependence of the components of a human system.

INTRODUCTION

E. Gary Stickel

      Artifacts and ecofacts are the primary "facts" of archaeology. Facts, however, do not speak for themselves; they must be interpreted. Interpretations are not made in a theoretical vacuum; all are based on theory in one form or another. As Henri Poincaré has stated, "Science is built of facts the way a house is built of bricks, but an accumulation of facts is no more science than a pile of bricks is a house." Over the last hundred years, archaeologists have accumulated a great number of facts concerning all of the various times and places of human habitation. However, very little in the way of interpretive theory has been developed (cf. Read and LeBlanc 1978). The authors of the present volume share the view that systems theory can provide a useful framework within which to organize the facts and systematically interpret the world-wide archaeological record.

      The use of systems theory in archaeology has been advocated by many archaeologists (e.g., Binford 1965, 1968a; Clarke 1968; Watson, LeBlanc, and Redman 1971; Renfrew 1972; see Plog 1975 and Koerper 1976 for discussions of the historical development and use of systems theory in archaeology). However, its use has been recently and extensively criticized by Salmon (1978) in an article entitled "What Can Systems Theory Do for Archaeology?" Salmon seems to believe that systems theory has only added a new "flashy vocabulary" and "jargon" to the field of archaeology. Many of Salmon's criticisms are valid ones; however, she concludes with the following statement:

> I think it would be unreasonable for archaeologists to expect either
> general systems theory or mathematical systems theory to provide
> them with a developed theory that can be adapted to archaeology,
> or even with much specific help in constructing an archaeological
> theory [Salmon 1978:182].

Yet in the same article Salmon states that "we are all familiar with many types of systems . . . [including] sociocultural systems" (Salmon 1978); she thus tacitly admits that there are "systems" of culture which can be studied. In addition, with regard to the "new vocabulary" issue, Salmon (1978:182) also admits that "a new language does sometimes enable us to look at things in new ways" (see Drover in this volume). Despite the inconsistency of her statements, one is left with the distinct impression that Salmon would altogether drop the use of systems theory in archaeology as unfeasible and, hence, unproductive. She apparently would discard systems theory because it was improperly used by archaeologists in the past. However, disregarding theories because of past or occasional misuses is not

scientifically productive. Science is a self-correcting process, and
certainly archaeological scientists would be unable to correct a theory if
it were to be prematurely discarded.

I view systems theory as the potentially most productive theory
for archaeology at the present time (ecological theory may be subsumed under
the living-systems subset of systems theory). If archaeologists were to
reject systems theory altogether, they could not be expected to generate
much in the way of a meaningful or valid scientific understanding of human
cultural variability and of the processes which have caused cultural change.
The elimination of this viable theory might well lead to an unproductive
analytical era which future scholars of archaeological theory could justi-
fiably refer to as a time of "paradigms lost." The history of science in
general has involved a series of refinements in both theory and concepts,
a process which has led to new and useful knowledge. Archaeologists should
seize every opportunity to advance our knowledge of the past by refining
such useful theories. The present volume offers several new approaches to
applying systems theory to archaeological analysis.

Anthropologists and archaeologists have traditionally referred to
their basic unit of study as a "culture." Those using systems theory have
referred to it as a "cultural system" or "sociocultural system." In the
first paper presented in this volume, a systems model (referred to as a
"human systems" model) is proposed which is more refined than those pre-
viously suggested.[1] This model, which is certainly subject to revision
and refinement, is expressly formulated to facilitate archaeological
analysis. The human systems model justifies its use as an analytical tool
by emphasizing (among other points) the role of human biology and its interplay with human cultural development, something which earlier models of
cultural systems unfortunately failed to do. The study of such systemic
relationships will lead to a new and better defined knowledge of the past.

The second paper, by Christopher Drover, attempts to refine the
manner in which archaeologists systematically conceptualize their basic data
sets, and is an outgrowth of previous work by the author and myself on the
problem of the terms used for these basic data sets. Hence, the present
paper deals with the problem of how terms such as "artifact" and "ecofact"
might unnecessarily limit our understanding of the ways in which material
items relate to the operation of human systems. Such an examination of the
referents of the terms applied to the basic data sets of archaeology can
lead to a better understanding of the processes by which the data become
incorporated into, utilized by, and even discarded from human systems. An
important point to note here is that new insights into the ways in which
archaeological data reflect past human life will undoubtedly develop as
these data sets are used to test new hypotheses. It must be recognized that
the terminology used affects the way data are classified; this, in turn,
structures the "reality" which is perceived. Drover's paper also explores
the question of whether human systems are uniquely different from other
living systems.

---

[1]From my own perspective, the term "human systems" is more accu-
rate than other terms (e.g., "culture," "cultural system," or "sociocultural system") which have been used traditionally; therefore, I consistently use this term to the exclusion of others.

In the third paper, Henry Koerper and I present what we believe to be an especially productive perspective from which to view primary cultural processes. In this model, the primary cultural processes are viewed as analogs of the mechanisms of biological evolution: selection, mutation, gene flow, and sampling phenomena such as genetic drift. Specific definitions of these processes are proposed within the context of a unified genetic analog model which helps to explain cultural change. We suggest that valuable information on the dynamics of cultural stability and change will be generated if archaeologists study cultural processes from this new systemic view.

Next, Joseph Tainter presents an analysis of symbolism in mortuary practices and shows how it reflects the systemic operation of different cultural contexts in the Midwest, Central and Southern California, and Hawaii. Evidence is advanced that components (e.g., biological, material, and ideological components) are interlinked in each cultural context and are definable by proper analysis.

The final paper, by Michael Glassow, shows how the material elements of systems (which may be objectively defined as "facilities") can be shown to have functioned in the system of which they were a part. Using an archaeological example from the Southwestern United States, the author explains how an adequate understanding of the role of facilities will pro- vide archaeologists with an informative insight into the operation of past "cultural" systems.

It should be noted here that the authors of these papers are not advocating any one single approach to the use of systems theory in archaeology. However, a unifying theme in all of the papers is the belief that systems theory is a viable and useful tool for archaeologists because of the fact that its holistic paradigm stimulates and facilitates the search for *relationships* among data. If those relationships are understood, then individual facts can effectively be applied to archaeological analysis. The further productiveness of systems theory will be readily evident if the goal of refining its concepts and applications is retained and promoted. Systems theory may be viewed as a "key" to the "door" of understanding our human past--but it is a key that needs to be modified and refined before that door will be fully unlocked.

These papers were expressly solicited from the various authors with the specific goal of addressing systems theory in mind. All of the authors are grateful to Ms. Carolyn Kiefer, who was the format/syntax editor of the initial manuscript; we also wish to thank Thomas C. Blackburn and Lowell Bean of Ballena Press for their encouragement and constructive comments. Finally, I am personally indebted to Claudine Scoville, who kindly did the artwork for the cover.

1   A General Human Systems Model for Archaeological Analysis

E. Gary Stickel

<u>INTRODUCTION</u>

Modern archaeology is striving to make itself relevant to today's world. It can do so only if it can maximize our understanding of how humankind and its culture have developed. Such an understanding can be attained if archaeologists derive a maximum amount of information from the artifacts and other remains which comprise the archaeological record. Furthermore, the desired information must pertain to all major aspects of past human life if this comprehensive, *holistic*, scientific understanding of the evolution of human culture is to be achieved. A general human systems model (briefly outlined here) is presented in the context of these new views on the expanded potential of archaeological research.

Although an increasing number of highly useful specific models have already been published, relatively few general models have been proposed for archaeological research. Moreover, these models pertain only to "culture" per se. This reflects the fact that the major unit of analysis (the one which archaeologists were trying to reconstruct and understand) was traditionally "culture." The resultant traditional "archaeological culture" model was first successfully challenged by Binford (1962, 1968) and Clarke (1968), who proposed "cultural systems" models. However, their emphasis was still on *culture*, organized into various "systems." It is suggested here that those systematic advances need further modification. Thus, a new general human systems model for archaeological analysis is proposed. The model incorporates variables of both human culture (including material culture such as artifacts and non-material culture such as ideas) and human biology. It is important to note that the underlying basis of this *holistic* model is the conviction that archaeologists are capable of obtaining much more information from their data than they have in the past. This model should facilitate research into the nature of the various human systems

---

[1] I am indebted to Ms. Carolyn Kiefer who edited the text and worked it up on a word processor. I am grateful also to Ms. Renda Mishalany for her additional editing and typing. Thomas Levy kindly provided me with the illustrations. I also wish to acknowledge the inspiration I received from Professor Lewis Binford and Professor Colin Renfrew under whom I studied and who provided the impetus which motivated me to generate the present paper, although they have not reviewed this presentation. Finally, to my students who have read and commented upon the model, I extend my heartfelt thanks.

which have been developed at various times and places on earth. Such
research should provide insight into past system organization, structure,
content, function, and change, especially adaptive change.

A wide array of specific research models have been proposed by
archaeologists to date (e.g., Adams 1966; Binford 1968a, 1968b, 1972;
Clarke 1968, 1972a, 1972b; Clarke and Chapman 1978; Cleland 1976; Flannery
1972; Hill and Gunn 1977; Kramer 1979, Leone 1972; Longacre 1970b; Redman
1973; Redman et al. 1978; Renfrew 1972; Renfrew and Cooke 1978; Schiffer
1976, 1978; South 1977). Although the formulation of such specific models
is essential to archaeology as a science, general models are fundamentally
needed for three main reasons: (1) they allow existing research to be
organized into understandable contexts; (2) they indicate those research
areas which have not been adequately explored; and (3) they provide concepts
and terms which facilitate scientific understanding. Such models provide
for the scientific explanation of the development of past human systems.
They can also provide for the formulation of valid universal principles or
laws, which is certainly one of the major goals of archaeology (Binford
1972; Fritz and Plog 1970; Klein 1977; Read and LeBlanc 1978, 1979;
Stickel 1979; Stickel and Chartkoff 1973; Gould 1980:36-42).

PREVIOUS GENERAL MODELS

Until recently, archaeologists shared a concept of "culture" which
differed from that shared by other anthropologists, particularly cultural
anthropologists. Whether explicitly stated or not, these concepts of
"culture" served as general models for research. Because cultural anthropologists could observe living people, they could directly record behavioral
attributes such as language, ceremonies, religious beliefs, and other
aspects of human interaction. Consequently, their definitions of culture
tended to stress the so-called "non-material culture" of various peoples.

These past cultural anthropological concepts were mainly concerned
with what some modern cultural anthropologists would consider to be only a
part of the "total system" utilized by a given population of people. These
former definitions stressed such aspects as "techniques," "social structure,"
"complexes of ideas," and the "personalities" of people (Weiss 1972:1377);
it has been argued that much of the modern cultural anthropological view
is also "partialistic."

It is quite obvious from the literature that, of all these partial views of culture, the most popular--increasingly so--is the
restriction of culture to mentalistic phenomena, to ideas or the
like (non-material culture) in the minds of men [Weiss 1972:1377].

Weiss criticizes these views because they either minimize or neg-
lect altogether the role of artifacts and material culture in the operation
of culture. Weiss prefers to define culture in terms of ". . . certain
total systems which deserve to be called *cultural systems*" (Weiss 1972:
1387; emphasis added). His systems would include both material and non-
material cultural variables.

Whereas many cultural anthropologists tended to minimize the
study of artifacts and other types of material culture, archaeologists

held the view that they were necessarily limited to the study of material
culture--the ruling dictum being "archaeologists cannot dig up ideas, only
artifacts." This view was emphasized by the formulation of the concept of
an "archaeological culture," suggesting a necessarily different entity from
that studied by cultural anthropologists, as a consequence of the "nature
of the data." It is probably fair to say that V. Gordon Childe's definition
of an archaeological culture has been generally accepted in archaeology until
recently:

> An archaeological culture is an assemblage of artifacts that
> recurs repeatedly associated together in dwellings of the same
> kind and with burials of the same rite. The arbitrary peculi-
> arities of all cultural *traits* are assumed to be concrete expres-
> sions of the common social traditions that bind together a cul-
> ture. Artifacts hang together in assemblages, not only because
> they were used in the same age, but also because they were used
> by the same people, made or executed in accordance with tech-
> niques, rites or styles prescribed by social tradition, handed
> on by precept and example and modifiable in the same way [Childe
> 1957:51; emphasis added].

In following Childe's model, if distinctive associations of
artifacts were found together at a number of sites, then one could suggest
that an "archaeological culture" had been demarcated. As a result,
archaeologists focused their attention on the comparison of traits in the
form of artifacts and artifact styles, with the intention of defining the
extent of their presumed cultures in space and time. Since it was considered relatively easy to compare certain artifact forms which exhibited
much stylistic variation (such as painted designs on pottery), analytic
attention was further restricted to only a portion of material culture--
that comprised of decorated or stylistically-discernible variation on
artifacts. Also inherent in the reasoning of that time was the notion that
certain inferences could be "reasonably" made about the "cultures" represented by the artifact traits. And since bones, shells, or other floral
and faunal remains were often found in association with the excavated
artifacts, it was also considered reasonable to infer the subsistence
technology and economy of the culture under study. This led archaeologists
to place such a disproportionate emphasis on studies of economy that one
prehistorian was moved to playfully coin the term *Homo economicus* to describe
their unit of analysis (Clark 1972:15).

But what did these archaeologists think about the other elements
of culture? It was generally assumed that "idea systems" (e.g., social
organization, political institutions, and religious practices) did not find
adequate expression in the material remains of past cultures. Therefore,
scholars could not hope to reconstruct, study, or explain their presence
using archaeological data.

This simple view of prehistory (called the "limitation of
inference in archaeology"; Smith 1955) was successfully challenged by
Lewis R. Binford (1962, 1968:18-23), who condemned the invalid trait-list
method as the "normative approach." More importantly, he maintained that

> . . . the argument that archaeologists must limit their knowledge
> to features of material culture is open to serious question
> (including the supposed dichotomy between material and non-
> material aspects of culture itself) [Binford 1968:21].

Binford stressed that what was needed was a general model more
representative of the complex human phenomena which archaeologists were
actually attempting to study: ". . . culture is not additive and consists
of more than summed traits--we could argue further that *culture is a system
of interrelated components*" (Binford 1968:24; emphasis added).

The observable variations in archaeological data, according to
Binford, reflect different subsystems (components) in a *cultural system*,
and these data may be expected to vary independently in the normal operation of the system or during change in the system (Binford 1965:203).
Binford, stressing the "extrasomatic" (i.e., extragenetic) adaptive quality
of culture, also outlined the composition of a general cultural systems
model--a model which is shared by many archaeologists today:

> Prior to the initiation of such studies by archaeologists we must
> be able to distinguish those relevant artifactual elements within
> the total artifact assemblage which have their primary functional
> context in the (1) social, (2) technological, and (3) ideological
> subsystems of the total cultural system [Binford 1962:218; Fig.
> 1.1].

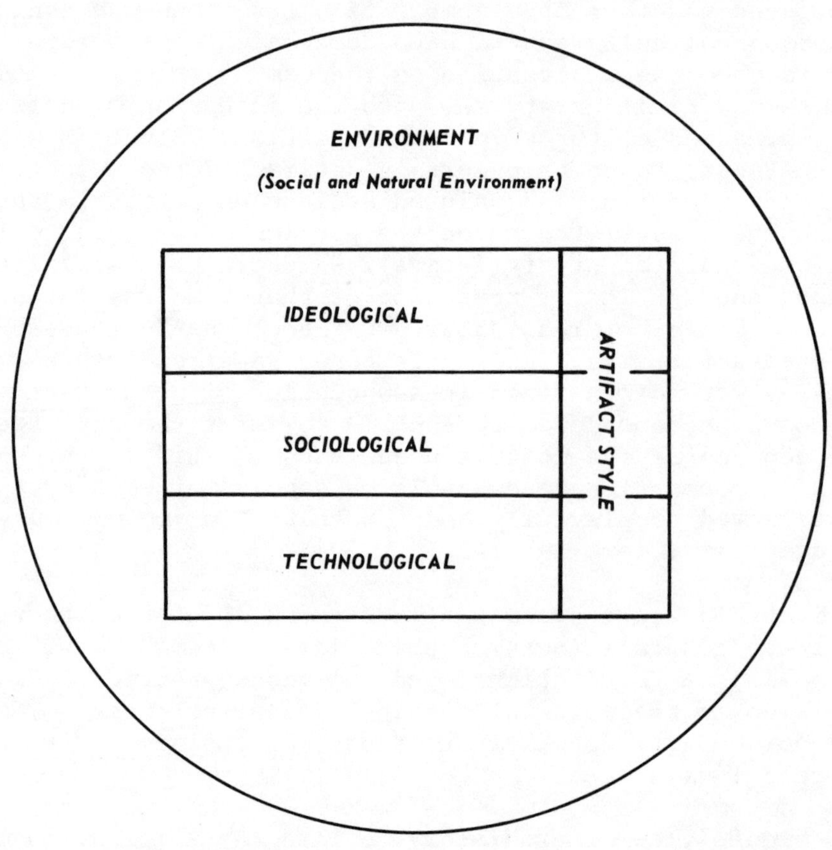

*Fig. 1.1. Composition and organization of a
cultural system (after Binford 1962, 1968a)*

In conjunction with other colleagues, Binford effectively promoted
this "systemic" approach to archaeology in the book *New Perspectives in
Archaeology* (Binford and Binford 1968). In the same year, another major
exponent of a systemic approach, David Clarke, presented a different view
of culture in his book *Analytical Archaeology*:

> The *sociocultural system* is an information transmission system
> of cumulatively acquired variety, supplementing instinctive
> behavior of man. The system institutionalizes and fossilizes
> evocation of particular stimuli, past and present, and trans-
> mits the unique code of the particular cultural experience of
> particular environments and partly idiosyncratic noise intro-
> ducing variety peculiar to the channels of that system. Such
> a sociocultural system is selectively advantageous for the
> survival of the individual [Clarke 1968:129].

Clarke expanded the number of suggested cultural subsystems to
five: the (1) social, (2) religious, (3) psychological, (4) economic,
and (5) material subsystems, respectively (Fig. 1.2). Clarke saw each
of these subsystems as affecting equilibrium conditions as a result of
feedback relationships between themselves and the total cultural system.
In turn, the total sociocultural system was seen as being in equilibrium
with its ". . . coupled environing system (i.e., the greater environment)
on an adaptive-selective basis" (Clarke 1968:126; Fig. 1.2).

More recently, Renfrew (1972) has also emphasized a systemic
definition of culture that stresses the causal interactions between a
cultural system and the environment with which it interacts:

> To represent the culture as a system or as part of a system, it
> is useful to consider not only the preserved artifacts, but the
> members of the society that produced them, the natural environ-
> ment they inhabited and the others (including the non-material
> ones such as language and projective systems) which they made or
> used [Renfrew 1972:19].

Renfrew suggests his own five-fold *cultural system* whose parts consist of
the (1) subsistence, (2) technological, (3) social, (4) projective or
symbolic, and (5) trade and communication subsystems. Renfrew does not
maintain that his particular cultural model is the only valid way of
organizing the data. He points out that the choice and number of cultural
subsystems recognized are arbitrary and depend upon the theoretical
orientation of the analyst (Renfrew 1972:20). This view is essentially
the one upon which the human systems model proposed here is based.

A HUMAN SYSTEMS MODEL

Although the earlier models discussed above are sophisticated,
they can no longer be considered completely adequate since they fail to
account for all essential system variability. It is important to note that
these cited "cultural systems" models do not adequately emphasize the role
of human beings. Humans are obviously the operators of the system; they
actuate culture. Without them there is no system. Hence, from this perspective, it is inappropriate to define a "cultural system" without
including the human population that is essential to the functioning of

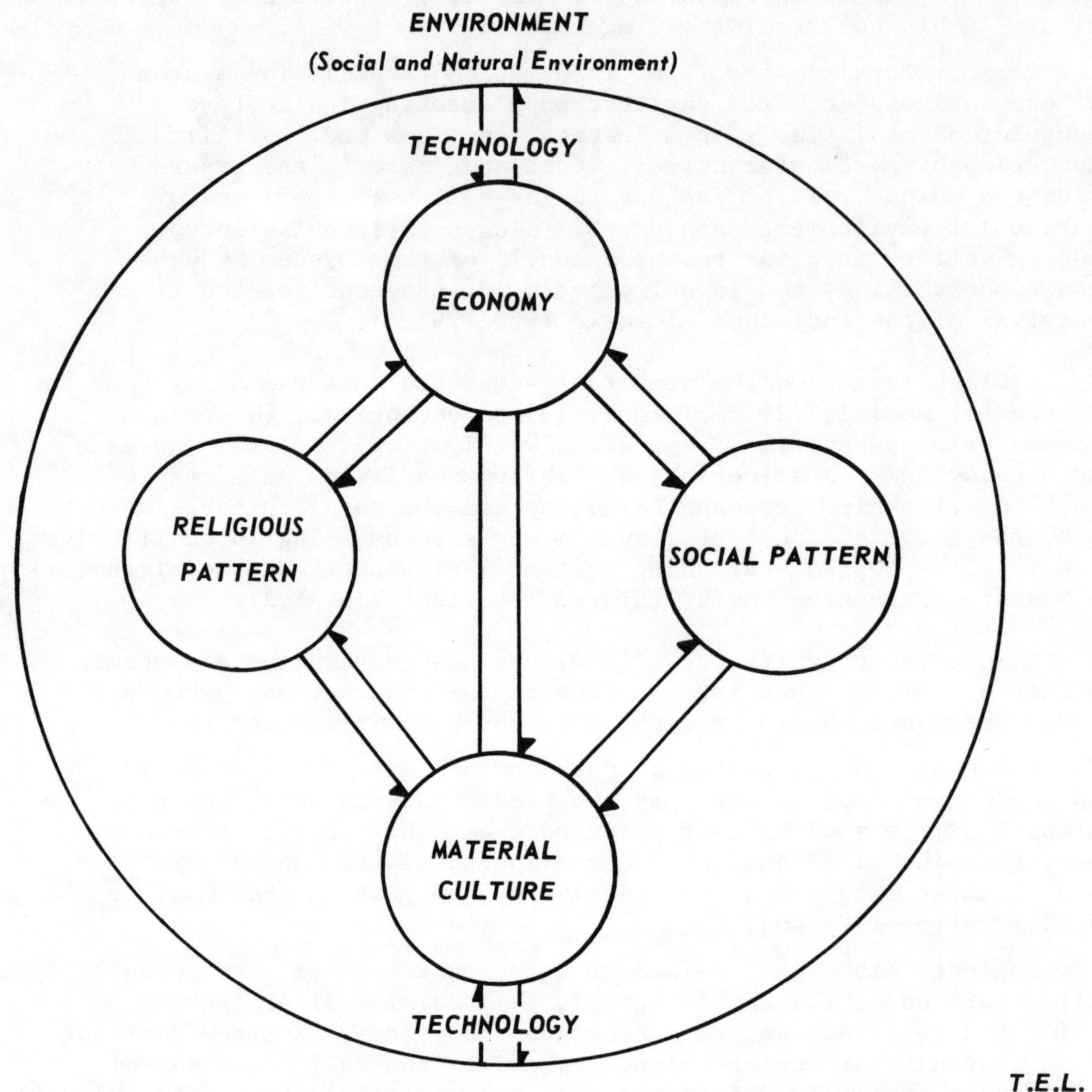

Fig. 1.2. Composition and organization of a cultural system (after Clarke [1968:103]. Clarke's Psychological Component is not depicted in this illustration. Nevertheless, it is discussed in his book [1968:113-114])

the given system. Moreover, human beings are not cultural per se; they
should be considered to be primary biological organisms comprising a
separate component in the system. Therefore, it is suggested here that the
term *human system* be used to refer to that system which a given population
of humans utilizes in day-to-day living. This term is designed to replace
what have been referred to traditionally as "cultures," "cultural systems,"
and "sociocultural systems." Also, by referring to the systemic phenomena
as a human system, one can eliminate the paradox of including human beings
(i.e., biological organisms) as elements in a "cultural system," since
genetics, not culture, produces human beings. The term human system also
refers to a simplified model of reality devised to demonstrate certain
organizational and operational principles. Thus a *human system* is a construct primarily comprised of culture (both material and non-material) in
conjunction with the biological organisms which actuate culture--a population
of people. All humans in a given system interact with one another, sharing
ideas and other material/non-material cultural elements (such as artifacts),
in a *total environment* within which their system exists (see Koerper and
Stickel this volume).

A general systemic model must not be overly simplified in such a
way that important system variability is neglected. Nor should such a model
be so complex as to prohibit either experimental control or realistic
operationalization with tangible data. In addition, the most salient aspects
of a proposed systemic model must be its identified subsystems; a useful
general model must identify subsystems which can be universally recognized
within all human systems. For instance, J. G. D. Clark's cultural model
recognizes a component of "war" (Clark 1972:15), although war is not a
universal cultural phenomenon. It is suggested that the human systems model
proposed here satisfies all these criteria.

In the present analysis, the term *system* will refer to an entire
human system. A distinctive human system is one which has more total
variation between itself and other human systems than it does within itself.
One could, for example, identify an American Southwest Navajo human system
versus a Hopi human system, or a BaBira human system versus a Pigmy human
system in the Ituri Forest of Zaire. Since all human systems are "open
systems" (see below), they are bound to share some variability with other
human systems (especially today, with increased communication, transport,
trade, etc.). Nevertheless, a distinctive human system may be identified
by means of various probability measures of association and interaction
(Barth 1969; Berry 1967; Hodder and Orton 1976; Spoehr 1973; Wobst 1974).
However, it is explicitly recognized that it will be a difficult task to
isolate some human systems from others.

<u>OUTLINE OF THE COMPOSITION AND ORGANIZATION</u>
<u>OF A HUMAN SYSTEM</u>

Systems theory, as it pertains to archaeology and culture, has
been discussed favorably by several archaeologists (Clarke 1968:43-130;
Hole and Heizer 1973; LeBlanc and Redman 1971; Plog 1975; Read and LeBlanc
1978; Renfrew 1972), even though Salmon (1978) has recently expressed some
scepticism regarding the analytical utility of systems theory. However, a
brief recapitulation of certain premises in systems theory (based on Miller
1965a) should be presented in order to emphasize the nature of human systems.

First, any system in general is an organized whole comprised of a ". . .
set of units with relationships among them" (Miller 1965a:200, 1965b;
Odum 1963:Chapter 2). All systems fall within one of two categories--
living or non-living. Human systems are living systems (Odum 1975) and
possess the following attributes.

(1) They are open systems (matter, energy, and information can
be exchanged with their total environments).

(2) As long as they exist, they try to maintain a state which
counteracts entropy (referred to by Miller 1965a as "negentropy")--i.e.,
they maintain order, organization, patterning of their elements and components,
and have nonrandomness of organization. When living systems fail to maintain
negentropy they disintegrate and perish. During that process, they leave
behind structured remains--structured in terms of the form and spatial
distribution of systematic data. This latter aspect is of prime importance
to archaeological studies of human systems; since once-living human systems
were ordered, organized, patterned, and had nonrandom organizational aspects,
and since matter (in the form of human/non-human bones, artifacts, and the
like) were incorporated into these systems, it follows that a defunct system
will leave behind structured non-random remains. These remains, in turn,
provide an opportunity for the archaeological interpretation of the nature,
structure, function, and operation of once-living human systems (Hodder and
Orton 1976; also see Schiffer 1976 for the complexities of how such remains
may be distributed or redistributed due to natural and/or cultural factors).

(3) Living systems ". . . have more than a certain minimum degree
of complexity" (Miller 1965a), and human systems may be considered the most
complex systems ever developed on earth. Moreover, most past human systems
have tended to evolve towards increasing complexity, as evidenced in the
archaeological record.

(4) As living systems, human systems contain genetic material
which is contained within their human organisms.

(5) Living systems have *deciders* which control the entire system.
Deciders in human systems are the human beings who consciously or unknowingly
make decisions about the operation of the system, especially its cultural
aspects.

(6) Living systems contain certain critical subsystems (components) whose presence and mutual interaction produce and maintain the
organizational whole--in this case the entire human system. Living system
"wholes" are more than a sum of their parts, due to their functions/processes.
These functions involve matter, information, and energy flows, both within a
system and between a given system and the total environment in which it
operates.

It is suggested in this paper that all human systems may be
considered to have at least seven critical components, including the (1)
biological, (2) material, (3) techno-economic, (4) sociological, (5)
ideological, (6) psychological, and (7) communicational subsystems (Fig.
1.3). Each of these components shall be discussed separately below. These
components are integrated into human systems to form an ". . . actively
self-regulating, developing, reproducing unitary system, with purposes and

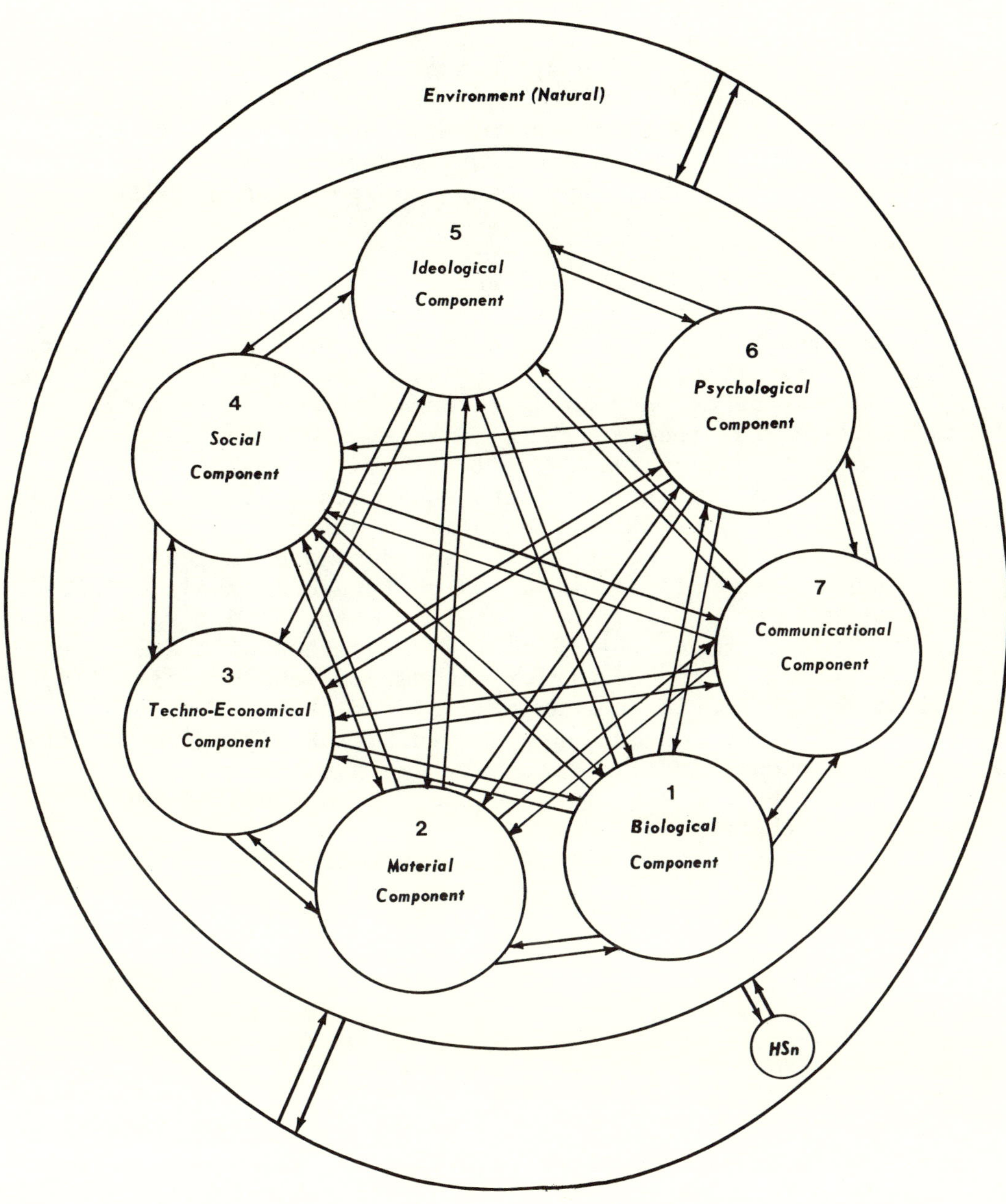

KEY

Human System: Biological, Material Cultural and Non Material Cultural Variables.

Biological Variables — Component 1   Material cultural variables — Component 2

Non Material Variables — Components 3–7

Human System : (HSn)

Environment: 1) HSn : Social Environment   (all other HS which interact with a given Human System)

2) Natural Environment : Climate, Geology, Flora, Fauna, etc.

T.E.L.

Fig. 1.3. Proposed composition and organization of a *human system*

goals" (Miller 1965a:204). (The foremost of these goals is survival.) Even
though seven components are recognized here, it is important to emphasize
that the phenomena which comprise human systems are exceptionally complex.
A system is actually comprised of a large, but probably finite, number of
variables (*n* dimensions of interacting variables). Hence, the number of
components recognized here involves an arbitrary reduction of reality to
facilitate a modeling of the entire composition and operation of human
systems.

Since culture comprises a significant amount of the variability
in a human system, it is necessary to define the concept operationally.
Weiss (1972:1396) suggests that culture can be defined as "the generic term
for any and all human nongenetic, or metabiological phenomena." In the human
systems model, culture may be defined as all those nongenetic (or extragenetic) related elements and processes in a human system which include
(1) all organized *information* (ideas), (2) all cultural *means* (processes) of
system operation, and (3) all *media* (artifacts and ecofacts). The cultural
differences and similarities between various human systems correspond to
the degree to which those systems possess similar ideas and express these
ideas via their cultural means of operation and their use of media. Ideas
are of critical relevance for a system when they are expressed through means
and media. In this context, *means* refers to the phenomenon of the linkage
of ideas (information which is the result of the perception, organization,
storage, and evaluation of data in the human mind or elsewhere) to behavior
(observable activity, actions, conduct, and nongenetic/physiological bodily
responses to input of various kinds). Means are amply illustrated in the
archaeological literature by references to methods, techniques, procedures,
and "human behavior" in general, as inferred from archaeological data.
Within this context, *media* refers to all of the material properties of a
human system that are represented by artifacts and ecofacts (Binford 1965;
see Drover this volume). Media have been frequently described in the
literature as "artifacts," "faunal and floral remains," and the like. All
human populations use certain means and media in order to insure the survival of their distinctive human systems. Archaeology can reconstruct and
explain the means of operation of past human systems through the analysis
of representative samples of preserved media (see Muller 1975).

It is important to recognize that a human system, like any living
system, is able to effect certain feedback relationships which operate to
maintain the system in a relatively unchanged state (homeostasis) within
its environment. Or, if changes are forced, the system is regulated in such
a way that the induced changes do not cause irreparable harm to the functioning of the system as a unitary whole. Archaeologists have pointed out that
there are a number of different possible feedback relationships. It has
been assumed that feedback relationships occur both between the component
parts of the system and between the system and its total environment
(Clarke 1968:Chapter 2). Since there are feedback relationships between
all components in a human system, it is suggested that these relationships
help to produce the precise variability of the material components of the
system. Therefore, by studying relevant variability in artifacts and other
media, archaeologists should be able to reconstruct aspects of all components
of once-living systems (Fig. 1.3).

Figure 1.3 shows the model with its system components graphically
represented by equal-sized spheres. However, it is not suggested here that

all of the components within each given human system are necessarily equal
in importance. The relative importance of the various components in a
system must be analytically determined by future research. It is antici-
pated that the components will vary within each human system and/or over
time.

All human systems exist, or existed, within total environments,
whether perceived as such by a given system's operators or not (Rodgers
1969). This total environment consists of a *natural environment* and a
human environment. The natural environment includes all relevant living
non-human floral and faunal systems, as well as non-living systems such as
water, geology, topography, climate, and other phenomena that have an effect
upon the physical features of the earth (Clark 1968:126). The human environment includes all other human systems which interact with a given human
system. If analysts are to understand environmental adaptations, they must
be concerned with the "effective environment" of a given human system; i.e.,
"those parts of the total environment which are in regular or cyclical
articulation with the unit under study" (Binford 1968:323).

All human systems have developed their specific systemic components in a manner that enables them to maintain themselves in their environments for the various periods of their existence; in other words, they have
adapted themselves to their environments. It is important to realize,
however, that no human system can determine *a priori* what sorts of adaptations will enhance survival, since a system cannot be aware of all past,
present, or future conditions relevant to that survival. This is why certain
human systems emerged at the close of the Pleistocene at the expense of other
hominid systems whose components were not as successfully organized. It is
one of the major goals of archaeology to reconstruct and understand the
adaptations of past human systems.

The Biological Component

> Selection has been applied to the family and not to the individual,
> for the sake of gaining a serviceable end.
>
> —Charles Darwin

> The most important advance in biology will be the discovery of the
> ways in which human heredity affects cultural evolution and cul-
> ture affects genetic evolution.
>
> —E. O. Wilson

The biological component consists of the total population of
mutually interacting and interdependent humans who are the operators of a
given human system. It should be pointed out that the total population consists of the aggregate of all population subunits of a distinctive human
system; for example, the combined population of all the villages once occupied by a distinct American Indian tribe. People in a distinguishable population are the organisms which make it a living system, and they are the
necessary actuators of the system; the system ceases to exist as a distinct
human system without this population. Thus even if it were possible to
completely replace the biological component of a particular human system
while maintaining the cultural components, it would no longer be valid to
consider it the same human system. A new system would have to be visualized,

since the nature of the biological variation within any given system is here considered essential to its definition as a distinct human system.

A system's biological component may be described in terms of its population genetics, anatomy (e.g., by anthropometry), demography (population structure, birth rate, death rate, and other similar aspects), human ecology (biologically speaking), evolution (e.g., by paleontological studies), and characteristics of physiology, growth, health, and/or pathology (e.g., Weiss 1973). The continuing study of "race" emphasizes the fact that different human systems have different biological ranges of variation. It is certain that many neighboring human systems have overlapping variation. It is not suggested in the human systems model that human biological variability will discretely co-vary with a human system. However, it is postulated here that a given system can be partially defined by the fact that more *inter-* than *intra-*systemic human biological variability should be present. The precise nature of a given human system's biological component is dependent primarily upon genetic processes such as gene mutation, gene drift, gene flow, and natural selection.

Genetic processes generate a range of potential phenotypic variations, which are then realized via precise articulations of organisms with their relevant environments. For instance, Lee (1972) points out that human biology can vary between populations of essentially identical genetic composition due to such factors as mobility and the nature of the food being consumed by a population (an example, focusing upon the !Kung system of Africa, is discussed later in this paper).

The study of variability in the biological component in human systems and its generation is generally considered the province of physical anthropology. But it is the task of the archaeologist to assess and explain the nature of the interaction between the cultural components (both material and non-material) and biological components of past human systems (Bray 1972; Spooner 1972). (The numerous cases of intentional cranial deformation practiced around the world are only one example of many such interrelationships between culture and human biology.) However, it is suggested that the operation of a system's means and media has fewer obvious impacts (Campbell 1976). No adequate understanding of the operation of past human systems can be achieved until these aspects are reasonably comprehended.

It is generally assumed that human biology evolved in concert with culture throughout the Pleistocene. However, the existence of anatomically modern man during the Holocene has led many researchers to assume that human biological variation during this period has been either non-existent or of insignificant proportions; therefore, they have reasoned, human biological variation is not important to archaeologists who are concerned with recent cultural evolution. The view taken here is that the role of human biological variation vis-à-vis cultural variation has not been sufficiently recognized by archaeologists. The human systems model suggests that all other components potentially are reflected in the biological component. Human biological variation provides for cultural variation, although perhaps not to the causal extent maintained by the biologist Edmund Wilson in his controversial book *Sociobiology: The New Synthesis* (1975). The basic theoretical position taken in the human systems model is that human genetics provides the potential for culture; however, the precise forms which comprise culture are determined by complex interactions both within a total system and between

that system and its total environment. While genetics provides for the
range of human behavior, it does not precisely determine variability in
cultural behavior (e.g., ideas and artifacts). Wilson's thesis may be termed
biological or genetic reductionism; in contrast, the human systems approach
is consistent with modern anthropological views. However, much of anthropological and, certainly, archaeological research has altogether ignored
the indispensable role of human genetics (the major determinant of the
biological component) in the operation of human systems. This situation
must be rectified if archaeologists are to conduct research that will achieve
a unified understanding of the operation of and changes in human systems
through time.

There is increasing evidence that human evolution has scarcely been
static since the onset of biologically modern populations. One example of
the continuing interplay between human biological evolution and culture can
be seen in work carried out by Kretchmer, who suggests that one food product,
cow's milk, which generally has been considered by western nutritionists to
be a universal human food, cannot in fact be used as a source of nutrition
by many adults in several contemporary human systems because ". . . the normal
genetic state of adult man is one which cannot digest milk due to a lack of
the enzyme lactase which is needed to break down the milk sugar lactose"
(Kretchmer 1972). Kretchmer hypothesizes that those human systems which have
developed lactase have derived it from prehistoric chance mutations in
ancestral populations which had utilized dairying:

> A person who could not digest lactose might have had difficulty
> in a society that ingested non-fermented milk or milk products,
> but the lactose-tolerant individual was more adaptable: he could
> survive perfectly well in either a milk-drinking or a non-milk
> drinking society [Kretchmer 1972].

Kretchmer also suggests that this mutation, through natural selection, might
have become a viable part of human populations around 10,000 years ago, when
dairying was in its early stages of development. Although the dating may be
questionable, this hypothesis points out the relationships between culture
and the biological component in human systems. In this instance, new techno-
economic *means* (animal husbandry) led to newly exploited *media* (dairy-
related artifacts and the actual milk products which were utilized as human
food). This food was taken from animals which were initially exploited for
other reasons, and was then used by humans and processed for their consumption. Those human-system biological components (individuals) who developed
this genetic capability (digestion of milk products) could increasingly
exploit the new resource; this in turn helped to promote the survival of
those respective human systems.

The archaeologist can excavate data which directly represent the
biological component of past human systems. These data are in the form of
human remains, which include preserved whole human bodies or portions of
them, tissues, skeletons, and various human waste products such as fecal
remains (coprolites).

Innovative analyses of human biological remains can not only provide archaeology with information on aspects of past subsistence practices,
they can illuminate the relationships of human biological components to all
other components in those past human systems as well. Relevant complexes of
human biological traits may be studied to provide such insights (Weiss 1973).

For instance, one innovative study ("Osteology of Social Organization:
Residence Pattern"; Lane and Sublett 1972) utilized non-metric osteological
traits in a human burial population (biological component data) to reconstruct marital residence patterns (sociological component data).

It is important to recognize that cultural components help the
biological component to survive. Although one archaeologist has stated
that "such a sociocultural system is selectively advantageous for the survival of the individual" (Clarke 1968:129)--a fact that Darwin recognized
long ago--it is the population of organisms, *not* the individual, which forms
the critical unit of survival for life forms. Therefore, archaeologists
must study the adaptation of each whole population that comprises a distinctive human system if a thorough understanding of human cultural evolution is to be achieved.

The Material Component

Any understanding of social and cultural change is impossible without a knowledge of the way media work. . . . All media are exten-
sions of some human faculty--psychic or physical. The wheel is
an extension of the foot. The book is an extension of the eye.

---Marshall McLuhan

The material component consists of all those non-human items,
objects, liquids, gases, substances, flora and fauna (especially domesticated ones), and similar data, which are or were directly incorporated into
a human system. These are things that are directly or indirectly (e.g.,
by animals or machines) handled, transported, built, utilized, or consumed by
the biological component of the system. All the elements of this component
may be termed *media*. For archaeological analysis, this media set contains
two major classes of elements, *ecofacts* and *artifacts*. More specific
definitions of these terms are provided elsewhere, but for present purposes
the term *ecofact* is restricted to those material things, taken from the
environment, which were directly handled and/or used by human beings in a
human system. Ecofacts are natural objects or substances which have *not*
been intentionally or purposefully modified into preconceived or specified
forms--forms which are then directly used for some purpose(s). For instance,
if a shell were to be discarded at a site after the shellfish was removed
for food, the shell would be considered an *ecofact*. However, if that shell
was subsequently modified by human activity so that it could be strung on a
necklace as an ornament, it would then be an *artifact*. Artifacts form the
other major class of *media* in the material component of a human system.
Media which have been fabricated or manufactured into intended, specified,
and/or utilized forms are artifacts. Artifacts (tools, ornaments, facilities, and other media) and ecofacts constitute life-sustaining media and
are therefore critical to the operation of human systems.

Artifacts, of course, have always been emphasized in archaeological analyses. In the human systems model, artifacts reflect (in terms
of specifiable complexes of variables) aspects of all components of the
human system of which they are (or were) a part. However, most archaeologists thus far have only studied techno-economics via the analysis of artifacts.

Because of the interlocking relationships between the material
component and all other components in a system (Fig. 1.3), the material
component's total media (depending on such variables as differing human
disposal practices, and natural or cultural disturbance and preservation)
provide a data set with which archaeologists can potentially reconstruct
and explain aspects of the entire componential organization of a past human
system. Indeed, it is necessary to reconstruct aspects of the means of
operation of all human system components if an adequate understanding of
system variability and change is to be achieved.

The biological components of human systems may be partially reconstructed through the study of burials, human dentition, human body-waste
products, and other such data. But the remainder of a human system, its
cultural components, should be reconstructed through the relevant analytical
study of media.

Binford was the first to propose that artifacts were complex items,
each of which could be related to a different component in a cultural system
(Binford 1962). He hypothesized that different classes of artifacts would
function separately in each of the three components in his "cultural systems" model (Fig. 1.1), and suggested the following terminology for artifacts
(it should be understood that once an archaeologist isolated each class of
artifact, the subsystem to which those artifacts were related might then be
reconstructed and studied):

> *Technomic* signifies those *artifacts* having their primary func-
> tional context in coping directly with the physical environment.
>
> . . . *sociotechnic*. These *artifacts* were the material elements
> having their primary functional context in the social sub-
> systems of the total cultural system.
>
> . . . *ideotechnic*. Items of this class have their primary func-
> tional context in the ideological component of the social sys-
> tem [Binford 1962:219].

I consider this terminology useful to archaeology because it
emphasizes the fact that different aspects of a system's components are
reflected in artifact variability; however, it must be modified and expanded
in the manner suggested below. Binford implied that *entire artifacts* may
be exclusively *related to just one system component*; for instance, he suggested that an axehead might be exclusively related to his "technomic"
component. In the human systems model, however, any given system medium,
such as an artifact, potentially reflects all human system components
simultaneously. This view suggests that certain *complexes of variables* in
an artifact (rather than the entire artifact) may be related *to particular
system components*. Therefore, it would not be appropriate to refer to a
"technomic artifact" per se, but it would be appropriate to refer to the
technomically-related variables in an artifact. It is suggested here that
the suffix *-technic* be used to refer to the relevant means of operation of
the system (e.g., sociotechnic, ideotechnic, etc.). The suffix *-fact* is
used in the present analysis to refer to the actual material variables which
have their primary functional contexts associated with different system components (see Table 1.1). An artifact represents a tangible nexus of systematic variability. Given this theoretical orientation, an archaeologist
can analyze an artifact into different complexes of variables, with each

TABLE 1.1

BREAKDOWN OF A HUMAN SYSTEM'S COMPONENTS, COMPONENT MEANS OF OPERATION AND REPRESENTATIVE MEDIA VIA THE ANALYSIS OF A KING'S CROWN*

System Components	Component Related Media ('Facts')	System Means of Operation ('Technics')
1. Biological	Biofacts (direct material variables which consist of biological data, such as a skull, or indirect indications of biological variables shown by artifactual media): No direct biofacts, such as a human cranium from a wearer of the crown, are present. Instead, the size of the headband of the crown is indicative of the cranial variability within the human populations which utilized the crown for use by selected persons of the highest status.	Biotechnic functions (the genetic/physiologic means of system operation): Genetic-physiological system variability is not directly represented by the crown, but indirectly indicated in terms of the human crania which could accommodate the crown.
2. Material	Materiofacts (non-human variables of artifacts/ecofacts found in the system): These are represented in the example by the materials of gold and gems used to create the crown.	Materiotechnic functions (the means of the functioning mechanical, structural, physical/chemical properties of the system's media): In this instance, they are variables of the materials of the crown which are specially used for its general purpose, such as the structural properties of gold as it can be used in metal headgear.
3. Techno-economic	Technofacts (media variables indicative of a system's technological and economic level of operation): In the example, they are those variables that exhibit the means of the crown's production such as the kind of gold alloy used.	Technomic functions (the means of all technological/economic operation of the system): In the example, this refers to the technological means of the production of the crown such as gold metallurgy and the craft of jewelry making.
4. Sociological	Sociofacts (material variables in the system's media that denote aspects of social organization): The crown's unique form is used as a symbol of an individual's sovereignty and serves to identify the ruler.	Sociotechnic functions (the means of the operation of the system's social organization): In this instance, this refers to the means of the exercise of the social power and influence of the crown's wearer.
5. Ideological	Ideofacts (distinctive material variables which exemplify a portion of the system's composite ideology): For example, the presence of jewels on the crown and the golden crown itself represent the use of information relative to the acquisition, modification/production and utilization of gemstones and gold.	Ideotechnic functions (the means of ideological operation/utilization of system information): All information (ideas) associated with the creation and use of the crown. In this example, ideotechnic functions would include the organized information relative to acquiring, modifying/producing and utilizing gemstones and gold.

| 6. Psychological | <u>Psychofacts</u> (media variables which indicate the system's psychology): Most often these are exhibited by attributes referred to as *style*. The distinct form of the cross attached to the crown, the precise shape of the gems and their placement on the crown and the specific artistic fashion of the depicted scenes from the Old Testament all represent the system's psychology of design. | <u>Psychotechnic functions</u> (the means of operation of the system's psychology): In this case, this variability pertains to the system's aesthetics involved in the decorative design of the crown, or the psychological impact a crown's wearer has on his observers in the system. |
| 7. Communicational | <u>Communifacts</u> (system's media which are variables indicative of communication within the system): In this example, the engraved words 'PER ME REGES REGNANT' on the crown exemplify a certain type of human communication, generally a form of symbolic inscribed language; specifically, here, the written language of Latin. | <u>Communitechnic functions</u> (the means of operation of the human directed communication in the system): This is exemplified by the usage here of a written language--specifically, Latin, on the artifact. |

---

[*](cf. Binford 1962; the particular crown analyzed is the Nuremburg Crown, Hugh-Jones 1963:98-99). The terms "technomic," "sociotechnic" and "ideotechnic" were proposed by Binford (1962). The remaining terms are proposed and defined here in order to facilitate a human systems analysis in a parallel fashion. It is maintained that this king's crown represents more than just "sociotechnic functions" as suggested by Binford (1962). Indeed, it represents all components of the human system of which it was a functioning element. Research which has been directed to reconstruct each of these components is cited in the concluding section of this paper.

complex reflective of that component to which it was related. This approach
is particularly feasible in light of recent advances in sophisticated artifact typology (Read 1974; Christenson and Read 1977).

If analyzed properly, the material variability in an artifact can
provide information on the various components of a human system, as a consequence of the causal feedback relationships between all components within a
given system (Fig. 1.3). Such feedback relationships provide for complexes
of variables in artifacts indicative of different components. Binford used
a specific artifact to exemplify his own model; this may be reanalyzed from
the perspective of the human systems model. Thus Binford suggested that a
"king's crown" was a "sociotechnic" artifact that reflected the social
organization of a system (Binford 1962:219). But from the perspective of the
human systems model, a king's crown would not be exclusively sociotechnic.
It certainly would be partly (or perhaps primarily) sociotechnic, in that it
would have been used by the paramount operator of the *sociological component*
in the system. However, a crown may be related simultaneously to *all* other
system components (see Table 1.1). For instance, the head-band size of the
crown would be related to human cranial variability in the system's *biological
component*. Materials in the crown would represent variability in the *material
component*. The presence of exotic precious jewels, as well as the gold
metallurgy involved in the production of the crown, would be indicative of
variability in the *techno-economic component* of the system reflecting means
of acquiring and producing such items. All of the relevant information
required to construct and use the crown would reflect the *ideological component*. The precise artistic styling of the crown would be referable to
aesthetic aspects of the *psychological component* of the system. The form of
communication represented by the writing on the crown (Latin script) could be
related to the *communicational component* of the system. Thus, from the viewpoint of this type of analysis, a king's crown represents *all* system components.

Artifacts represent multiple dimensions of human system operation.
Once an analyst has isolated relevant complexes of variables in artifacts,
aspects of the means of operation of one or more system components may be
partially reconstructed. Although an artifact may potentially reflect all
system components, any given artifact may not function equally in all components. For example, the *primary* function of an axe may be techno-economic
(i.e., to chop wood for fuel or construction purposes); however, from the
perspective of the human systems model, the axe could still reflect all of
the other system components to lesser and varying degrees through the
presence of correlating variables in the axe (such as design elements on
the axe indicative of the social group to which the axe's user belonged).
If media that comprise the material component of a past human system are
properly analyzed, archaeologists can reconstruct aspects of the componential
structure of that entire system. Such material culture analyses will facilitate explanations of a system's variability and changes through time, and
thereby add to our understanding of the evolution of human systems in
general.

<u>The Techno-economic Component</u>

> Production produces not only an object for the subject but a subject for the object.
>
> --Karl Marx

The definition of a human system's techno-economic component involves a number of concepts. David Clarke (1968:103) defined his "material cultural subsystem" in his general model as "the patterned constellations of artefacts which outline the behaviour patterns of the system as a whole and embody that system's *technology*" (emphasis added). He next conceived of his "economic subsystem" as "the integrated strategy of component *subsistence methods* and *extraction processes* which feed and equip the society" (Clarke 1968:103; emphasis added). As was noted above in the discussion of the material component, Binford distinguished three general functional classes of artifacts reflective of the three subsystems in his model: (1) technomic, (2) sociotechnic, and (3) ideotechnic artifacts (Binford 1962:219); the "technomic" subsystem comes the closest to conforming to the present model's techno-economic component. Binford states that those artifacts which reflected his technomic subsystem primarily functioned in coping with the natural environment. Like Clarke, Binford emphasized subsistence and extraction processes. He also defined technology as ". . . those tools and social relationships which articulate the organism (cultural system) with the physical environment . . ." (Binford 1962:218). Although artifacts do "embody" a system's technology (as in Clarke's quote on the "material cultural subsystem" above), they also potentially reflect all other components of a human system as well. However--technology here is considered a major part of a separate system component--*the techno-economic component*. The techno-economic component of a human system consists of all the techno-logical/economic phenomena in that system. Technology, in this model, refers to the technical means of a system's acquisition and usage of matter, energy, and information. Technology involves more than the human-directed procurement and transport of raw materials (termed "extraction processes" by Clarke and Binford); it also includes all means of procuring, transport-ing, processing, fabricating/manufacturing, constructing, using, revising, maintaining/repairing, and finally disposing of or abandoning the media acquired by a given human system. All of these phenomena are closely involved with a system's techno-economic component. Technology also involves work--human or human-directed work via the use of animals or machines. Work demands an outlay of energy, in the form of either human or non-human energy expenditures.

However, from the standpoint of the human systems model, tech-nology does not refer to the "tools and social relationships" suggested by Binford. It is true that technological activity usually requires both tools and social relationships (e.g., miner work-groups with their equipment and supervisors). However, in a human systems analysis, tools per se are *media* of the material component, and social relationships are the social *means* of operation of the sociological component. In order to perform techno-economic activities, humans in a system draw upon the required means and media in the other components to accomplish needed tasks.

Technology and economics are considered to be conterminous in the human systems model. All technological acts are simultaneously economic acts, since they involve the acquisition and transmission of matter, energy,

and information. Technology does "equip" a human system and allow for the
system's survival by providing the critical life support of the biological
component (in the form of food stuffs and water or other liquids). However,
with the development of increasingly complex human systems, there has been
a corresponding increase in the number and complexity of phenomena such as
goods and services which obviously are not human food stuffs. These goods
and services are needed by human systems in order to function in a variety
of complex ways; economists generally refer to these as "commerce," "exchange,"
or "trade." In spite of their primary functions, these phenomena are
partially the products of system development and interaction between all
components of a human system. Therefore, an archaeological comprehension of
past techno-economics involves the study of those means of operation (termed
here "technomic," but distinct from the entire technomic subsystem proposed
by Binford) which were used by a system to provide needed services and
goods (media) required by the system in its daily operation. The variable
complexes in media (e.g., artifacts) which have their primary functional
context in the techno-economic means of system operation may be productively
studied in order to understand past techno-economics. For instance, one
technomic activity of a given system may have been hunting. In terms of
analysis, the particular techno-economic means of that system may be
reconstructed and understood by isolating those material variables in the
weapons which enabled game to be killed. Such variables might be the pierc-
ing point of an arrow, the tensile strength of an arrowhead, the power of a
bow, and/or other related variables. Research interest might also lie in
reconstructing the trading activities of a given system. Such economic means
may be reconstructed and understood by analyzing those material variables
in artifacts which enabled trading to have taken place (e.g., the forms of
standard artifacts which were exchanged in the past, such as the specific
forms of shell money used by early native Californians or the obsidian
exchanged in the prehistoric network of the Eastern Mediterranean). Techno-
economics provides information, energy, and matter utilized by or involved
with all other system components; hence, there is a techno-economic aspect
to all media such as artifacts and ecofacts used in or by a human system.
For archaeology, the task consists of analytically reversing the process by
reconstructing the means of operation of past human system techno-economics
through the study of relevant aspects of system media. Such reconstructions
will facilitate the explanation of past human techno-economic development and
change.

<u>The Sociological Component</u>

> An aesthetic artifact has been converted into a sociological arti-
> fact.
>
> --Norman Mailer on New York graffiti

        The sociological component of a human system consists of all the
sociological phenomena within the system that are commonly clustered under
the term "social organization." These phenomena are considered here to be
part of the sociological component, regardless of their duration. They involve the organization of the human members of a given system (the biological
component) into various groupings deemed appropriate for performing necessary activities (e.g., various economic, social, religious, and other activities). The sociological component partitions the biological component into
cognitively-recognized groupings of people. Social organization is needed
for the utilization and operation of all system information, means, and

media. All human systems would cease to exist were it not for the ongoing
means (here termed sociotechnics) of human social organization. Thus, the
sociological component provides the binding fabric of a human system.

Social organization orders relationships between people in such a
way that it allows or restricts human social intercourse. Deetz (1967) has
suggested that "human behavior" may be reconstructed on four distinct levels:
that of the (1) individual, (2) group, (3) community, and (4) society (the
last of these Deetz equates with the complete set of communities which comprise a culture; i.e., here the total set of biological components in a
given total population of a human system). Moreover, Deetz has suggested
that each level may be reconstructed from distinctive artifact groups. Since
it takes two or more persons to comprise a form of social organization, it
may be suggested that three of his four levels can productively be studied
to reconstruct the sociological means of system operation. In other words,
groups, communities, and societies are socially organized in different ways
which either permit or inhibit social intercourse among all the people in a
given human system. Each of these levels can be reconstructed and understood through the study of relevant groupings of artifact attribute sets
(complexes of variables).

Social anthropology has long categorized social organization in
differing ways. For instance, Service (1962) recognized three stages in the
evolution of social organization: (1) the band, (2) the tribe, and (3) the
chiefdom. Fried (1967) preferred to emphasize status arrangements in his
evolutionary framework: (1) the egalitarian society, (2) the ranked society,
(3) the stratified society, and (4) the political state. Many anthropologists
have emphasized the form of kinship structures (e.g., families, clans,
moieties, matrilineages, and patrilineages). In addition, forms of human
social organization not based on kinship, such as sodalities, political
organizations, warrior groups, clubs, and today's business corporations, have
also been studied. Regardless of the formulation used, the main purpose of
such studies has been to reveal and understand important relationships
between all human forms of social organization, both past and present.

It is significant to note that, in human systems, the sociological
component provides for the operators of all aspects of system operation.
For example, it allows for marriage and/or mating arrangements, so that offspring (who will eventually replenish the biological component of the system) may be forthcoming. In certain systems, it furnishes shamans or doctors
who care for the physical and mental health of the biological component.
The social organization supplies all of the various groupings of people
(minimally, male and female groups) which acquire the media that comprise
the material component of the system. The sociological component makes
provision for the techno-economic personnel (e.g., hunters/gatherers,
herders, farmers, traders, navigators, craftsmen, miners, contemporary
factory workers, or businessmen) of the system. The organization of the
system's social actuators (e.g., village headmen, clan leaders, chiefs,
kings, foremen, corporation executives, or presidents) is maintained.
Ideological actuators (e.g., priests, teachers, policemen, judges,
philosophers, or politicians) are defined. Additionally, psychological
practitioners are furnished (e.g., "medicine men," psychologists, psychi-
atrists, therapists, or counselors). The sociological component also furnishes communications specialists (such as village spokesmen, orators,
scribes, signalers, messengers, couriers, secretaries, translators, computer

and/or telegraph or telephone operators, and radio and/or television operators)
for the system. In short, the sociological component provides the organization for all component operations within any human system.

Binford (1962:219) suggested that certain artifacts were "material
elements having their primary functional context in the social sub-systems of
the total cultural system. . . . Artifacts such as a king's crown, a warrior's
coup stick, a copper from the Northwest Coast, etc., fall into this category."
A different view is suggested by the human systems model, in that *relevant
sets of variables in artifacts*, rather than entire artifacts themselves, may
be reliably related to the sociological component of a given system. Therefore, archaeologists can contribute to the explication and explanation of the
development of human social organization through the effective study of relevant complexes of variables in artifacts (Table 1.1).

The Ideological Component

Ideas can't wait, something must be done about them.

--Alfred North Whitehead

The fifth component of a human system is the ideological component.
Other archaeologists have already considered some of the variability which
comprises this component, although they have tended to be too restrictive in
their conceptions. For instance, Clarke had a separate "religious" component
in his "cultural system model":

> Religious subsystem. . . . Under the definition followed here this
> is a device to isolate the body of *information* forming the struc-
> ture of mutually adjusted *beliefs* relating to the supernatural,
> as expressed in a body of doctrine and a sequence of rituals
> which together interpret the environment to a society in terms of
> its own percepta [Clarke 1968:110-113; emphasis added].

Binford acknowledged the existence of an ideological component by
referring to "ideological artifacts," but he restricted it to the "ideological
rationalizations" of a cultural system:

> Items of this class have their primary functional context in the
> ideological component of the social system [i.e., his total cul-
> tural system]. These are items which signify and symbolize the
> ideological rationalizations for the social system and further
> provide the symbolic milieu in which individuals are encul-
> turated, a necessity if they are to take their place as func-
> tional participants in the social system. Such items as figures
> of deities, clan symbols, symbols of natural agencies, etc.,
> fall into this general category [Binford 1962:219-220].

Thus, by his examples, Binford seems to indicate that religious and other
beliefs form the ideological component of his cultural system.

In the human systems model, the ideological component consists
of *all* phenomena relative to *ideas* or *information* in a human system. Thus
ideology is considered here synonymous with the information--especially all
non-biological information--contained within the minds of the various individuals that comprise the biological component, and/or encoded into media

such as books, computers, astronomical monoliths, or other artifacts. The
ideological component may be considered to be the collective intelligence of
the system, consisting of those non-genetically acquired bits of information
which are also referred to as "ideas," "facts," "data," "learning," "lore,"
"bodies of information," and so on.

This information is cognitively stored in the minds of the people
in any human system. The resultant informational "pool" is constantly being
changed (by both deletions and additions) during the lifetime of the individuals, groups, and whole societies which operate given human systems.
Individuals and groups in the system have differential access to this information, whether they consciously acknowledge this fact or not. Ideology is
information organized into forms which include those traditionally recognized
under such rubrics as religious beliefs or doctrines, magic, rituals, the
content of language, oral literature, music, laws, "designs for living,"
philosophies, scientific knowledge, histories, arts, humanities, and business concepts, to mention a few. In short, all non-genetic information
utilized in the operation of all relevant components in a human system comprises the ideological component. But it is important to recognize that the
ideological component deals with the information in a system but *not* with
human involvement with that information; a human system's ideology may include
a certain type of religion, but the extent and intensity of human belief in
that religion involves another component--the psychological component.

The information in a human system is organized in forms which
facilitate its deployment to relevant aspects of the system (e.g., techno-
economic information utilized in a technomic activity such as hunting, or
information necessary for sociotechnic activities--like how to properly
behave toward other individuals within a community involving various statuses). This non-genetic information was acquired and constantly changed by
means of human observation, study, communication, and learning. Its continued existence is determined by the human system of which it is a part--by
the survivability of the entire system (with its pool of information) and by
the applicability of the information towards system survival. Systems, in
turn, may vary in their ability to accept new information while retaining old
information; it is also important to note that some information is adaptive,
some is neutral, and some is maladaptive in terms of the survival of an entire
human system.

The ideological component is not considered here the passive
recipient of the total system. As Gerald Weiss (1972) asks, "Can't ideas
generate ideas?" The answer is a qualified yes; it is not ideas which
generate ideas, but our human ability to recognize creative possibilities
(a human psychological quality involving a partial awareness of the potential applications of presently available information) that allows new information to be included in a system. Some information may be acquired by a
system that is not fully capable of interpreting it; our attempt to understand archaeological data would be an example. Nonetheless, this information can be potentially decoded and understood by humans if they use the
various means and media at their disposal.

Analysts must consider the ideological component extremely
important to human systems because it consists of the basic, essential
information for system operation. It also provides the ideas, concepts,

goals, rationales, options, and motives for human action per se (whether
they are utilized or not depends on other factors, such as whether the
society will permit or prevent the implementation of these rationales for
behavior). Therefore, archaeologists must gain some kind of understanding
of the variability in the ideological component of a human system (its pool
of information) before they can understand the operation of the total system
under examination. Archaeologists can reconstruct the ideological component
of past human systems by studying those material clusters of variables in
artifacts which have their primary functional context in the ideological
component. The failure of some past human systems to take certain courses
of adaptive action in the wake of selective pressures might have been due
to the fact that those systems simply lacked the relevant information necessary to implement those adaptive actions.

The Psychological Component

> Governments rarely take note of this sentiment. . . . It plays no
> part in statistical tables; and historians ignore it, because it
> is one of those impalpable social phenomena. . . . The casual
> observer, however, easily sees the differences in the (psycho-
> logical) behavior of different societies. . . .
>
> --Jean-Francois Revel, *Ni Marx, Ni Jesus*, 1970

The above quote notes that psychological variability between
different human systems has not been properly recognized. There has been
a general bias against the analytic study of system-wide human psychology,
a bias that has also carried over into archaeology. As a consequence, the
ability of archaeologists to reconstruct and explain past human psychological variability has been discounted (Binford 1972:198). Nevertheless,
the viewpoint adopted here is that scientific insight into the operation of
the psychological component in human systems is mandatory if archaeologists
are ever to accurately understand the total operation of and changes in past
human systems. If suitable explanatory models are formulated and tested,
archaeology can then produce useful information on past psychology and its
causal role in human system operation and change (cf. Binford 1968a:23).

David Clarke (1968:102, 113-114) included a "psychological subsystem" in his formulation of a cultural system; he also developed a concept
of "group psychology" for archaeological analysis. A modification of these
concepts is adopted in the human systems model proposed here. While the
psychological component consists of all psychological variability within a
given human system, the sometimes highly idiosyncratic variation of individuals must not be emphasized at the expense of the salient psychological
features of a human system. These salient features are phenomena which
involve the systematic, patterned human deployment of feelings, emotions,
aggressions, sentiments, values, "tastes," creative inspirations, and
aesthetics vis-à-vis other humans, the various means and media of a given
human system, and parts of the total environment of the system as well. The
patriotic aggression that is mobilized during a war is a dramatic example of
such system deployment. The sometimes intense collective feeling engendered
by a religious ceremony would be a salient, patterned psychological feature
of the component, as are the feelings and evocations created by certain
media such as idols. Psychological processes involve (to varying degrees)
all means and media of a human system. Thus societies can be aware of

certain ideologies (e.g., capitalism or communism) without having strong
feelings about them, but some human systems can obviously operate with
respect to those ideologies with very strong value deployments.

Each human system develops its own psychological means of operation, which leads to the creation of clusters of variables in artifacts whose
primary functional context is associated with the psychological component.
For example, the worship of Buddha in India constitutes a form of patterned
psychology. This has led to the selection of certain variables in artifact
design and style decoration which, in this case, distinguish a given statue
as a highly recognizable representation of Buddha within the total media of
a given human system in India.

The psychological component is not considered here an unimportant
and passive source of variability in a system. It can account for the continued presence of certain system means and media in situations which at first
glance appear incomprehensible. A system's adherence to a form of adaptation
which proves maladaptive in the long run may be due to that system's psychological commitment to that adaptation. The often-demonstrated "conservatism"
of design elements on pottery, which endure for relatively long periods of
time, may be explained as a consequence of a system's strong psychological
commitment to those design elements. The "logic" of adaptive decision-making
in the past may be understood if a given human system's psychological component is explored.

It is suggested here that material cultural variability can reflect the psychological component in archaeological data. For instance,
although he did not consider it a separate component, Binford (1962:330, Fig. 1.1;
cf. Dunnel 1978) indicated that "artifact style crosscuts" his "ideological,"
"sociological," and "technological" subsystems. The viewpoint taken here is
that artifact style crosscuts system components because it is representative
of the psychological component; this simultaneously reflects (and is
reflected by) the other components in a human system. Therefore, it is suggested in the present model that stylistic attributes or variables (such as
pottery designs, decorative designs on architecture, decor, and aesthetic
community planning) may be particularly useful in reconstructing past psychological variability within a system (see Renfrew 1972; Deetz 1973; Martin
and Plog 1973; Fritz 1978).

The degree of development of the psychological component may well
be the hallmark of the human condition. Our human psychological ability to
create--often called the "creative impulse"--to a great extent distinguishes
us from other species, for it has allowed us to change our systems far more
rapidly than other faunal species, species which are much more restrictively
dependent on genetic change. The human ability to create is made possible
by genetic factors within the biological component, but the relative
emphasis on and acceptance of creativity is a patterned psychological-
component phenomenon that varies from one human system to another. Human
creativity can infuse new ideas (information) into a human system. Such
"cultural mutations" (see Koerper and Stickel in this volume) can lead to
both *innovations* (such as the creation of a new means of hunting or a com-
putational method) and *inventions* (such as the devising of new media like
the bow and arrow or a new form of computer). This ability to create has
provided human systems with an adaptive potential in the form of new means

and media for system operation. Such variability has undoubtedly been of
great significance because of the adaptive alternatives it has provided to
systems that needed to find new ways of coping with selective environmental
pressures.

The Communications Component

> Here then is the hypothesis I want to advance. . . . I am suppos-
> ing that in every society the production of discourse is at once
> controlled, selected, organized, and redistributed according to
> a certain number of procedures whose role is to advert its powers
> and dangers to cope with chance events to evade its ponderous,
> awesome materiality.
>
> —Michel Foucault, *The Archaeology of Knowledge
> and the Discourse of Language*, 1972

The last component isolated here for analysis is the communicational component. This component involves the transfer of information from
the human members of a system to each other, and from humans to media and
perhaps back again (e.g., computer readouts). The component consists of
all forms of human communication, including language (spoken, written, or
signed), song, dance, and various forms of "body language." A system's means
of communication lead to the creation of variable clusters of artifacts which
have their primary functional context associated with the communicational
component of a given human system.

Some excellent examples of material cultural variables indicative
of this component are the various symbols of a writing system. It is
important to note that the communicational component interlinks all components
and elements of a human system into a coherent and responsive whole. Mutual
communication within a human system enables the effective continued functioning of the system. Archaeologists have not suitably recognized the significant role communication plays in the operation and development of human systems, although communication has always been an essential part of those systems. Systems which could not adequately maintain internal communication
have perished more easily than those which could (the decline of the Roman
Empire, especially during its terminal stages, is a dramatic example of this).

This very brief description of the proposed human system components provides for productive anthropological analysis. However, no model
can be considered seriously until its utility has been demonstrated. Therefore, data from the African !Kung Bushmen are presented below in order to
illustrate the utility of the model as an analytical device.

AN EXAMPLE OF THE APPLICATION OF THE MODEL

The following section involves a human systems analysis of the
ethnographically-described !Kung Bushmen; although the chosen example is
ethnographic, its archaeological implications should be apparent. The !Kung
system has recently undergone significant change. (It is important to recognize that it may be analytically easier to identify and understand the componential and interactional aspects of human systems if a system is studied
which has undergone rapid and substantive change.) A relatively recent

presentation of data on these people provides an excellent insight into the organization of and changes in all the components of the !Kung human system (Lee and Devore 1972; also see Lee 1972; Kolata 1974; Yellen 1977).

The !Kung system will serve here as the given unit of study. The total environment of this system includes the natural environmental component of the Kalahari Desert of South Africa, as well as a human environmental component primarily comprised of Bantu neighbors. It is maintained here that all of the observed systemic changes in the !Kung system are due to the mutual relationships (feedback) between the !Kung system's components within the total human system and between that total system and its greater environment.

Kolata (1974) emphasizes the techno-economic component of this human system: "The !Kung have lived as *hunters* and *gatherers* in the Kalahari Desert of South Africa for at least 11,000 years; but recently they have begun to live in *agrarian* villages near those of the Bantus" (Kolata 1974: 932; emphasis added). The nature of the changes in all !Kung human system components can easily be seen in Kolata's account, which discusses the way the !Kung are adapting to a new environment imposed primarily by the Bantu. Kolata's data are organized below in terms of the human systems model, component by component.

!Kung Material Component

The changes in food (media) in the !Kung system, from wild edibles (i.e., wild nuts, vegetables, and meat) to cultivated foods, have been noted above. Essentially, the consumption of a variety of small game (such as the nocturnal spring hare) and important plant foods (such as the mongongo nut) has given way to the use of grain and cow's milk. In addition, many other changes in the system's media are taking place. It has been noted that the formerly highly mobile !Kung now (in their new sedentary role) ". . . cannot easily pick up and leave . . ." (Kolata 1974:933). This is because the media of their nomadic system were extremely simple and transportable in contrast to their newly acquired agrarian media. For example, nomadic shelters were simple, easily-constructed brush huts. Sedentary !Kung have now acquired many new, not-easily transported artifacts, such as immobile permanent houses, which they can readily retain by virtue of the fact that they do not have to transport them.

!Kung Techno-economic Component

The !Kung are changing their techno-economic means in several ways as a consequence of the fact that they are substituting an agrarian subsistence base for their former hunter-gatherer subsistence base. It is suggested that their technology of artifact fabrication and use will be greatly modified or changed altogether through acculturation, with an addition of new agrarian-related media. They are also learning to desire and value trade items and other exotic media appreciated by their Bantu overlords; therefore, they are now participating in new techno-economic means of trade and exchange.

!Kung Sociological Component

The mobile !Kung had an egalitarian emphasis in their sociological component. However, the settled agrarian !Kung are modifying this component,

particularly with regard to the sexual division of prestige. The settled
!Kung women now have less prestige, less autonomy, and less ability to
directly influence group decisions than they did when they were part of a
mobile system. Kolata (1974) notes that one anthropologist has suggested
that the hunting and gathering !Kung women enjoyed higher social prestige
because of their substantial contribution to techno-economics, since the
women formerly gathered resources that contributed at least fifty percent of
the consumed food. According to this argument, this economic clout formerly
put women in a more influential position. The agrarian women now remain in
the villages preparing food, maintaining their shelter, and rearing their
children. By contrast, !Kung men are emulating the male-dominated Bantu
society. They conduct more exclusive techno-economic activities relating to
food procurement than do women. Specifically, the men are directly engaged
in herding and farming tasks under the direction of the Bantu. The women
are excluded from these activities. Kolata suggests that these factors are
inducing !Kung women to adopt an increasingly subordinate role.

Moreover, changes in child rearing practices are apparently pro-
moting age/sex "class" distinctions which contribute to the loss of egalitar-
ianism. Mobile !Kung had small band organizations so comprised that youth
groups (including play groups) contained children of both sexes and varied
ages, a situation which discouraged the development of the games or roles
based on age/sex distinctions which characterize non-egalitarian social
structures. The settled !Kung have youth groups comprised of individuals of
about the same age and sex. Boys are expected to help their fathers with
herding and farming tasks, but young girls are induced to remain in the villages and help their mothers with domestic chores. These factors are inducing a new social organization or sociological means of operation in the !Kung
system.

!Kung Ideological Component

New ideas (information) are being incorporated into the !Kung
ideological component. Such new information includes ideas relating to herding, farming, sedentary village lifeways, Bantu artifact usage, trade good
usage, and activities relating to food procurement, processing, cooking, and
consumption. Other new information pertains to communication (Bantu language),
games, rank, and roles. These newly incorporated information sets are changing the total ideology of the !Kung system.

!Kung Psychological Component

Kolata (1974:933) notes that the mobile !Kung promoted egalitar-
ianism by discouraging aggression among children. !Kung children formerly did
not play competitive games (a fact that is attributed to the varied age/sex
mixture of play groups, which presumably makes competitiveness more difficult). Moreover, hunting and gathering children were constantly watched by
adults of both sexes, who quickly stopped aggressive behavior between children.
Kolata also states that these ". . . children rarely observe aggressive
behavior among adults because the mobile !Kung have no way to deal with
physical aggression and consciously avoid it" (Kolata 1974:933). The fluid
band structure permitted dissatisfied or disputant families to simply leave
one band and join another. In this way conflict and aggression were miti-
gated by group fission and refusion. Sedentary !Kung now have so much

invested in their new homes, artifacts, and general lifeway that they
". . . cannot easily pick up and leave, [but] rely on their Bantu neighbors
to mediate disputes" (Kolata 1974:933). Thus, the !Kung psychological component is being modified in such a way that new relationships of dominance and
subordination are being formed between the sexes and between age groups. It
is suggested by the present model that the settled !Kung are developing a new
group psychology of values, needs, and wants, including a "pride of ownership."
These psychological changes will not only permit a non-egalitarian social
structure to develop and thrive, but will have an impact on all means and media
of the system as well. The changes in the psychological component will allow
other cultural changes to be both accepted and acknowledged.

!Kung Communicational Component

        The !Kung communicational component is undergoing marked change.
A major aspect of this change is the fact that, since the men are working
directly under Bantu men, they are learning the Bantu language. In turn, the
Bantus ". . . deal exclusively with the men" (Kolata 1974:933). Kolata suggests that this differential learning and usage of the Bantu language is pro-
moting the formation of male dominance within !Kung society. From the perspective of the human systems model, it is suggested that other aspects of
communication (such as "body language," dance, and song) will probably change
as well.

!Kung Biological Component

        The fact that a human system's biological component must be con-
sidered in any comprehensive analysis of system change is dramatically illustrated by the observed changes in the !Kung biological component. The feedback relationships between the cultural components and the biological component are also clear.

        Kolata notes that while the former population size of the !Kung was
low and highly stable, a fact that reflected a hunter-gatherer demography set
in the Pleistocene, it today numbers 30,000, and the ". . . !Kung population
is rapidly increasing rather than remaining stable" (Kolata 1974:932).

        One remarkable systemic feedback link is having a profound impact
on the !Kung biological component--the diet (i.e., food items representative
of media from the material component). The varied, nutritionally-balanced
hunter-gatherer diet has given way to an agrarian diet comprised of much grain
and cow's milk. Kolata (1974:933) states that the "healthy" diet of the
mobile !Kung lacked vitamin deficiencies and many people were free from such
common diseases of old age as increased blood pressure. Studies carried out
over an entire generation indicate that the agrarian, settled !Kung are taller,
fatter, and heavier than they were as hunting and gathering nomads. In addition, it has been found that the agrarian women begin menstruation earlier
than the hunting and gathering women did. The increased time of fertility,
coupled with an observed decrease in the average time between births, is
leading to such marked demographic changes as a dramatic increase in population. The hunting and gathering !Kung women have "no soft food" to give
their babies (Kolata 1974:934); consequently, they nurse them three or four
years. During that time mothers rarely conceive. By contrast, agrarian !Kung
wean their babies much sooner, because they can give them soft food (media of
the material component) in the form of porridge and cow's milk; such weaning

affects fertility. Moreover, the cultivated foods have led to an increase in
the body fat of !Kung women; this, in turn, has affected both the early onset
of menstruation and fertility (ovulation). New techno-economic means (agricultural techniques) have provided new material resources (media in the form
of agricultural products) which have affected the !Kung system's reproductive
capability; this has then fed back on the system by providing a net increase
in population (the structure of the biological component is changed with the
appearance of new consumers and producers). Here one can see the feedback
links between the techno-economic, material, and biological components.

<u>Summary of the !Kung Human System</u>

In summary, it appears that all components of the !Kung human system
have undergone change. The mutual interaction (feedback) between components
will undoubtedly produce more system change as an adaptive response to continuing disruptive environmental factors. This process will result in a !Kung
human system at a new level of structural and functional organization.

If future archaeology were to be conducted at !Kung agrarian sites
and comparisons made with former !Kung hunter-gatherer sites, a number of
observable differences should be indicative of this systemic change. Future
archaeologists might find evidence of human skeletons, other human body products, and system-related media (ecofacts of animal bones and plant remains in
addition to the artifacts) with which to reconstruct the ways in which the
system had changed. For instance, changes in the biological component could
be reflected in taller stature in measured human skeletal remains. The change
to an agricultural diet might be reconstructed through physical/chemical
analyses of the teeth and bones, and by possible coprolite analysis. Diet,
nutrition, health, and system genetics could be inferred from dentition, precise bone morphologies, and by the survivorship curves characteristic of
sequent burial populations.

Changes in the material component would be well documented by the
presence of artifacts related to farming and herding, trade goods, and more
substantial house remains (mobile !Kung sites retain little if any trace of
brush huts). Also, the presence of ecofacts of domestic animal remains (cattle
bones) and domestic plant remains (perhaps recoverable through pollen analysis)
could indicate material component changes as well.

Changes in the techno-economic component would be reflected in an
observable change from artifacts related to hunting and gathering to those
associated with farming and herding. Bones reflective of a hunter-gatherer
technology (e.g., bones of the nocturnal spring hare; Kolata 1974:932) would
be replaced by cattle bones and plant remains indicative of species utilized
in the agricultural economy. The increased techno-economics of trade would
be indicated by the presence of exotic trade items less likely to be found
at mobile !Kung sites.

Evidence of changes in the sociological component might include
greater qualitative and quantitative differences between age and sex groupings
as reflected in grave-good associations and/or the treatment and spatial
organization of burials. Even the forms of houses and other structures, as
well as their spatial organization within the sites, could provide relevant
information.

Changes in the ideological component might be deduced from a much greater variability in artifacts and other media found in sedentary as compared to mobile !Kung sites (e.g., the presence of Bantu and other alien artifacts and more trade goods in the sedentary than in the mobile !Kung sites).

Changes in the psychological component could possibly be detected by comparing the limited number of design elements found on all artifacts at mobile !Kung sites with the greater variety of design and decorative elements to be expected at sedentary !Kung sites. (The latter would be due to the fact that the sedentary !Kung can both retain and require more stylistically-varied artifacts due to increasing social differentiation than the mobile !Kung.) In addition, there should be a much more complex and formalized village design and layout at a sedentary site than at a mobile !Kung site. The interior décor of houses should reflect changes in this component as well.

Finally, changes in the communicational component might be reflected in the presence of artifacts with written language or symbols on them from various other systems (i.e., trade items with trade marks). In addition, the !Kung might adopt or copy in their own idiom the design elements of the Bantu, thereby indicating an awareness of that other communication system.

In short, future archaeologists should be able to identify aspects of all the componential changes in the !Kung human system over time by the careful analysis of a representative sample of data from relevant sites.

SUMMARY AND CONCLUSIONS

A general human-systems model has been briefly outlined. A human system is analyzed into seven components considered reflective of essential systemic operation: the biological, material, techno-economic, sociological, ideological, psychological, and communicational components. It should be noted that the model considers as relevant for archaeological analysis all three sets of component variables, including *biological* variables (component 1), *material cultural* variables (component 2), and *non-material cultural* variables (components 3 through 7). Each of these components has complex feedback relationships with all other components. As a consequence, it is possible to reconstruct *aspects* of all components in a given human system by studying tangible remains in the archaeological record (e.g., human skeletal remains and artifacts). This is feasible since precise variation in that data is the result of the simultaneous causal interaction of all the components in a human system. Moreover, the variability within a human system also partially reflects variability in the greater natural and human environments to which it has adapted, and to which it will have to further adapt in the future as required by new environmental changes.

The present model may be subsumed under the general rubric of "systems theory." Salmon (1978) has criticized the archaeological use of systems theory and has made a number of valid observations with regard to problems with the theory. Despite these difficulties, the view presented here is that a "systems approach" forces an archaeological analyst to search for relationships and cultural processes in data instead of simply studying *entities* such as a non-functional stone-tool type. In spite of Salmon's criticisms, a number of archaeologists continue to find the systems approach

analytically productive (Read 1978; Redman et al. 1978; Renfrew and Cooke
1978). Modern archaeology is capable of such systemic analysis, especially
in light of its recent emphasis on multi-disciplinary projects. Current
research is already illuminating many of the identified human system components, as well as their various links to other components. For example,
the biological component has been explored recently in terms of archaeology
(Bray 1972; Gummerman 1971; cf. Spooner 1972; Weiss 1973), and some of the
links between it and other components such as the sociological sub-system
have been clarified (Brown 1971; Lane and Sublette 1972). The communicational component (Lieberman et al. 1972) has also been investigated recently.
The material component and its media have been well studied in archaeology,
of course, but that component has also been specifically related to other
components; e.g., plant remains and animal remains have been related to both
the techno-economic components of past systems (Brothwell and Higgs 1970;
Bokonyi 1974; Renfrew 1973) and to trade or exchange aspects of systems
(Earle and Ericson 1977); Sabloff and Lamberg-Karlovsky 1975). However,
archaeology has been disproportionately concerned until now with the study of
artifacts for typological and technological reasons, for the multiplex nature
of human artifacts is becoming increasingly apparent through current research.
For instance, artifacts (material component media) are being related to the
sociological component (Deetz 1967; Friedrich 1970; Hill 1970, 1976, Hill and
Gunn 1977; Longacre 1970a, 1970b; Renfrew 1975); to the techno-economic component (Krupp 1977; Renfrew 1972; Wobst 1976a, 1976b); to the psychological component (Deetz 1973; Martin and Plog 1973:113; Fritz 1978); and to the communicational component of past human systems (e.g., Chadwich 1961; Renfrew 1972;
Wobst 1976b). These studies, the various archaeological research summarized
in Pfeiffer (1977), and the analysis of the changes in the !Kung human system
in this paper all indicate that the human systems model can be effectively
operationalized.

It may be seen that the underlying thesis promoted throughout this
discussion is that archaeology can learn much more from its data than it is
presently accomplishing. That view is tied to another salient point, which is
that archaeologists are *not* limited to studying only certain aspects of past
human systems (such as technology and economy), as some researchers have long
maintained (Smith 1955; cf. Willey 1974). On the contrary, it should now be
apparent that archaeology has barely realized its potential to reconstruct
and explain evolutionary changes in the entire range of componential variation in past human systems. But "to achieve this requires a rejection of all
notions concerning the limitation of inferences" in archaeology to just
economy and technology, as Michael Rowlands (1978:327) has so aptly stated.
Hence, the human systems model is predicated on a view which is "open-minded"
about the potential for archaeological inquiry into past system variability.
I fully endorse the view made over a decade ago that

> propositions concerning any realm of culture--technology, social
> organization, psychology, philosophy, etc.--for which arguments
> of relevance can be offered are as sound as the history of
> hypotheses confirmation [in science]. The practical limitations
> on our knowledge of the past are not inherent in the nature of
> the archaeological record; the limitations lie in our metho-
> dological naivete . . . [Binford 1968:22-23].

The methodology which can best establish the validity of the
holistic approach is that of the iterative, deductive-inductive,

multiple-hypothesis testing procedure used in science in general (cf. the
procedural testing chart in Stickel and Chartkoff 1973:669; see Raab 1973,
1977; Schieffer 1976). The validity of archaeological inferences (or
interpretations) must depend upon the ability of hypotheses to withstand the
test of continued, objective scientific investigation; they must not be based
upon the intuitive, untested opinions of individual scholars. A continuing
emphasis must be placed on formulating new hypotheses, models, or theories,
if archaeology is to truly achieve its goal of understanding and explaining
past human-related phenomena. From this perspective, if archaeology is to
make meaningful contributions to science, it must fully adopt a *holistic*
methodological testing approach. This approach must be aimed at understanding
the complex interworkings of all the critical biological and cultural components of human systems and how those components, in turn, are affected by
a system's ecological adaptation to its total natural and human environments
over time. Given our uncertain future as a species, a knowledge of how past
human systems have adapted, changed, survived, or perished may help humankind's future survival through a better understanding of the required
"balances" in the interrelationships between system components and their
larger environment (see Martin and Plog 1973:361-368). The variety of past
and present human systems presents a truly wonderful panorama. At the very
minimum, this more holistic systemic approach will provide archaeology with
a powerful interpretive tool--a tool which will enable archaeologists to gain
a more complete understanding of the operation of and changes in these
remarkable human systems through the analysis of their preserved cultural
media. Given this perspective, and the immense amount of data representative
of the long span of human development on this planet, it should be possible
to formulate valid principles or laws of human system structure, function, and
change--a goal essential to any science seeking understanding (Schiffer 1976;
cf. Read and LeBlac 1978, 1978; Stickel 1979). Therefore, from the perspective of the model proposed here, archaeology should be defined as the scientific recovery, reconstruction, and explanation of the structure, operation,
and change of past human systems via the analysis of the preserved remains
comprising the archaeological record left by those systems.

2 Views on the Basic Data of Archaeology

Christopher E. Drover

In recent years, the working definition of "culture" has changed for archaeologists. For many, the concept of culture once referred to an assemblage of material objects representative of particular peoples at certain times in certain places; from these material objects, other aspects of culture (such as religion, social organization, and behavior) could be inferred (Meighan 1966:191). This particularistic view focused on the material components of culture, which resulted in reconstructions of historical sequences and lifeways. Such a conceptualization of culture, however, places certain cognitive limitations on explanations of culture change.

More recently, culture has been modeled as a human system in which the interrelated components are in a constant state of interaction or feedback (Binford 1962; Clarke 1968; Renfrew 1972; Stickel in this volume). Although the potentialities of this type of model have not yet been fully realized, it undoubtedly encourages a more holistic approach to problem-solving; this, in turn, should lead to a better understanding of cultural changes and processes. Indeed, general systems theory may allow archaeologists to deal with more than just "culture change" problems derived from rather general notions ("laws") about cultural evolution. From the perspective of systems theory, the distribution of material objects is thought to result from the patterning of all components of a system (e.g., social organization, ideology, or psychology). Each of the components of culture may vary independently, resulting in multifaceted causal mechanisms. Explanations of culture change need no longer be couched in terms of a single mechanism such as diffusion, but can be seen more realistically as involving a causal nexus of internal and external processes.

Recently, both methodological and theoretical changes have brought about the development of a new paradigm in American archaeology. Unfortunately, much of the terminology presently in use reflects the vocabulary (and hence the thinking) of an older normative concept of culture. Some of the critical terms presently in need of clarification are the subject of this paper. Such clarification should lead to terminology consistent with, and highly useful for, the systems approach in archaeology.

During past field experiences, discussions frequently arose with colleagues as to the proper categorization of materials present in a site; these perhaps had been modified through use, but had not been deliberately manufactured. Those discussions have resulted in the present consideration of the word *artifact*.

The term artifact has been variously used and defined in the past.
Little help is provided in traditional sources, such as the *Oxford Dictionary
of English Etymology*, in which we find the combination of Latin *ars* and *factum*
(*facere*--make, do), defined as ". . . product of human art" (Onions 1966:52).
This definition is not particularly enlightening because of its generality
and dependence on the concept of art. Other references give some idea of
the diversity of opinion that exists. Oakley (1959) defined artifacts simply
as ". . . objects deliberately shaped." This stood as a typical definition
of exclusively human behavior until Jane Goodall recognized tool-making among
chimpanzees in the 1960s (Oakley 1959). Since it was discovered that chimpanzees deliberately manufacture items, the definition has shifted to
essentially conform to Meighan's definition of the artifact as "any object
showing manufacture or use by humans" (1966:190). In more recent years,
authors have emphasized cultural norms, patterns, or templates as blue-prints
for artifact manufacture. Deetz, for example, has created an elaborate
analogy between the basic components of linguistics and those attributes which
make up artifacts (1967:83-95). He goes beyond analogy to suggest that

> . . . there may be structural units in artifacts which correspond
> to phonemes and morphemes in language . . . reflecting an essen-
> tial identity between language and objects in a structural sense.
> If this is true, in view of the close similarity between the way
> in which words and artifacts are created, might not words be but
> one aspect of a larger class of cultural products which includes
> all artifacts as well [1967:87]?

Similarly, Rouse suggests that an artifact is ". . . a structure, tool,
etc., modified in accordance with the norms of a culture" (1972:263). The
fact that the last two definitions seem restrictive regarding individual
creativity and over-emphasize cultural conformity are legitimate criticisms.
All of the foregoing definitions are inadequate upon closer inspection. In
the course of archaeological analysis, various items are encountered which
defy ready classification schemes; this difficulty emphasizes the short-
comings of the above definitions. For example, lithic debitage is variously
treated as either "artifactual" or "non-artifactual," as the researcher's
interest and interpretation of "deliberate" manufacture dictates. Lithic
debitage is analogous to the wood chips that result when a tree is chopped
down. In order to be systematically relevant, an artifact should be an
object that has been rendered into a specified form for specific use(s).
It is questionable whether all lithic flakes/wood chips fit the above criterion. This is not to say that non-artifactual material is not relevant,
only that it is relevant in a different way to human systems. For instance,
projectile points (artifactual) reflect different information than flake
waste (non-artifactual) with regard to human systems explication (cf. the
social organization studies involving projectile points [Deetz 1967] with the
more restricted inferences regarding technological-disposal patterns involving flake waste [Schiffer 1976]).

Although production is universally recognized as a property of
artifacts, it has not been sufficiently emphasized that artifacts may
result from use alone. This problem is exemplified by an ethnographic
observation made by McGee (1898) among the Seri. DiPeso's description of
McGee's observations concerning the Seri's use of an object which archaeologists might normally dismiss as simply a "rubbing stone" is especially relevant to the present discussion.

This naturally-shaped quartzite beach pebble had battering marks
about its edges, smooth rubbed surfaces on its flat planes, fire-
blackening on both sides, and flecks of red pigment on its surface.
In the course of a six-day period of observation, it was noted
that the stone was used (1) to skin the leg of a horse, (2) sever
the tendons of the leg, (3) to knock off the parboiled hoof of a
horse, (4) to crush and splinter the leg bone for marrow, (5) to
grind mesquite beans, (6) to pound shelled corn, (7) to chop a
small tree, (8) to break up branches for firewood, (9) to dethorn
octillo stems, (10) to sever a strong hair-cord, (11) to support
the bottom of a kettle while cooking over an open fire, (12) to
pulverize red face paint, and (13) occasionally to throw at
troops of dogs which entered the living area [DiPeso 1974:120].

It should be noted that the use of this stone is reflected, to some extent,
in the surface alterations, although purposeful human modification did not
occur *prior* to use.

The term artifact suggests, to many, an object modified prior to
or during a specific use. In this type of definition, debitage and undifferentiated but utilized flakes are excluded as not being artifacts *per se*
in the traditional sense. Human by-products in general should not be considered as artifacts, or else the definition becomes unmanageable. Artifactual items should be distinguished from non-intentional by-products of
cultural activity such as air pollution, temperature changes, defoliation,
sawdust, etc. Again, the relevancy of non-artifacts is not the issue, but
rather the fact that artifacts and by-products yield qualitatively and
quantitatively different information. In the present "archaeological-folk
taxonomy," a semantic dichotomy exists in which by-products must either be
artifactual or non-artifactual. It is therefore suggested that the concept
of artifact be revised to conform to a model based upon the idea of an energy-
exchange continuum. In this way, items like debitage, utilized flakes, wood
chips, air pollution, and tortillas can be separately viewed as material
expressions of *degrees of energy exchange*.

In order to contend with materials which are not artifactual (in
the sense that modification did not precede use) but are still related to human
systems, Binford has suggested the term "ecofact," defined as "culturally
relevant nonartifactual data" (Binford 1964). He goes on to imply that ecofacts are ". . . all those elements which represent or inform about the points
of articulation between the cultural system and other natural systems . . ."
(1964). Although Binford's term formally recognizes a wide range of relevant
data, its weakness lies in the fact that (in terms of systems theory) it
recognizes virtually no bounds. In the modern world, very little of the
natural environment is *not* an ecofact of human culture. The relationships
between soils, hydrology, pollen, and floral and faunal materials are inexorably
intertwined.

On a somewhat lesser scale, Mary Leakey has introduced the term
"manuport" to describe exotic lithic material (introduced to a site by hominids)
with little or no evidence of use (1971:3-4). While this term is a description
of one class of nonartifactual material, it is not suitable as a means of
referring to the myriad of other ecological materials that could relate to a
site (Binford 1964).

In addition to the inherent taxonomic problems presented by the word "artifact," the items of interest to archaeologists frequently belong to separate sampling universes. Binford (1964) has pointed out that, while the artifactual universe is defined spatially by the archaeological site, ecofacts may be sampled both from the site and from the region. It is therefore proposed that archaeological material information be comprised of three classes of data: (1) artifact, (2) ecofact, and (3) relevant environmental variables (REVs). These classes can then be revised to form energy continua. In this way, culturally-influenced materials such as debitage, utilized flakes, sawdust, air pollution, defoliation, and cultigens can be viewed in terms of the *degrees of energy exchange* which they represent.

Four modes of energy exchange can be considered in this definitional process: (1) procurement, (2) processing, (3) production, and (4) use/reuse. The *procurement mode* involves the initial collection and transportation of a raw material from its source to the archaeological site in question. Many materials in archaeological sites are recognized as being exotic to the area and show no modification beyond their natural state. The *processing mode* is a stage in which by-products are produced by modification, thereby making part of the raw material useful. It involves cultural data such as lithic debitage, sawdust, or any object produced by an activity for further use in a system; this category primarily involves ecofacts. Archaeologists must take care to identify natural vs. human modification in order to distinguish the mode of activity. The *production mode* involves an energy exchange threshold; the result is an object whose form has been modified for specific use(s). Although this implies the cognitive recognition of a desired finished form on the part of the maker, it may not be distinctively recognized (emics). As will be seen below, the production mode is a necessary component of the traditional definition of *artifact*. The *use/reuse mode* is a form of energy exchange that involves alterations resulting from human utilization. These alterations to an object may not be purposeful, in comparison to the production mode. Reuse here refers to the recycling of previously-used artifacts back into the cultural system through various "transformation systems" (Schiffer 1976:29).

Theoretically, each of the above energy transformation thresholds can be empirically measured through observable changes in the physical state of objects. The energy states described above need not be sequential; steps can be skipped (see Fig. 2.1). In this way, material objects that reflect cultural systems need not be confined to artifactual vs. nonartifactual, or artifactual vs. ecofactual, categories. Existing terms that refer to archaeological data can be better interpreted in the processual framework.

In the present model, an "artifact" must go through at least *two* energy exchanges, one of which involves procurement and the other either a use or a production mode. An "artifact" can, of course, pass through a processing mode (i.e., a retouched flake), but need not necessarily do so.

An "ecofact" has to go through only one energy exchange, the procurement mode (used by a system). Many archaeological "ecofacts" also go through a processing mode which renders part of the material useful to the human system, although the "ecofacts" themselves have *not* been a part of production or use modes.

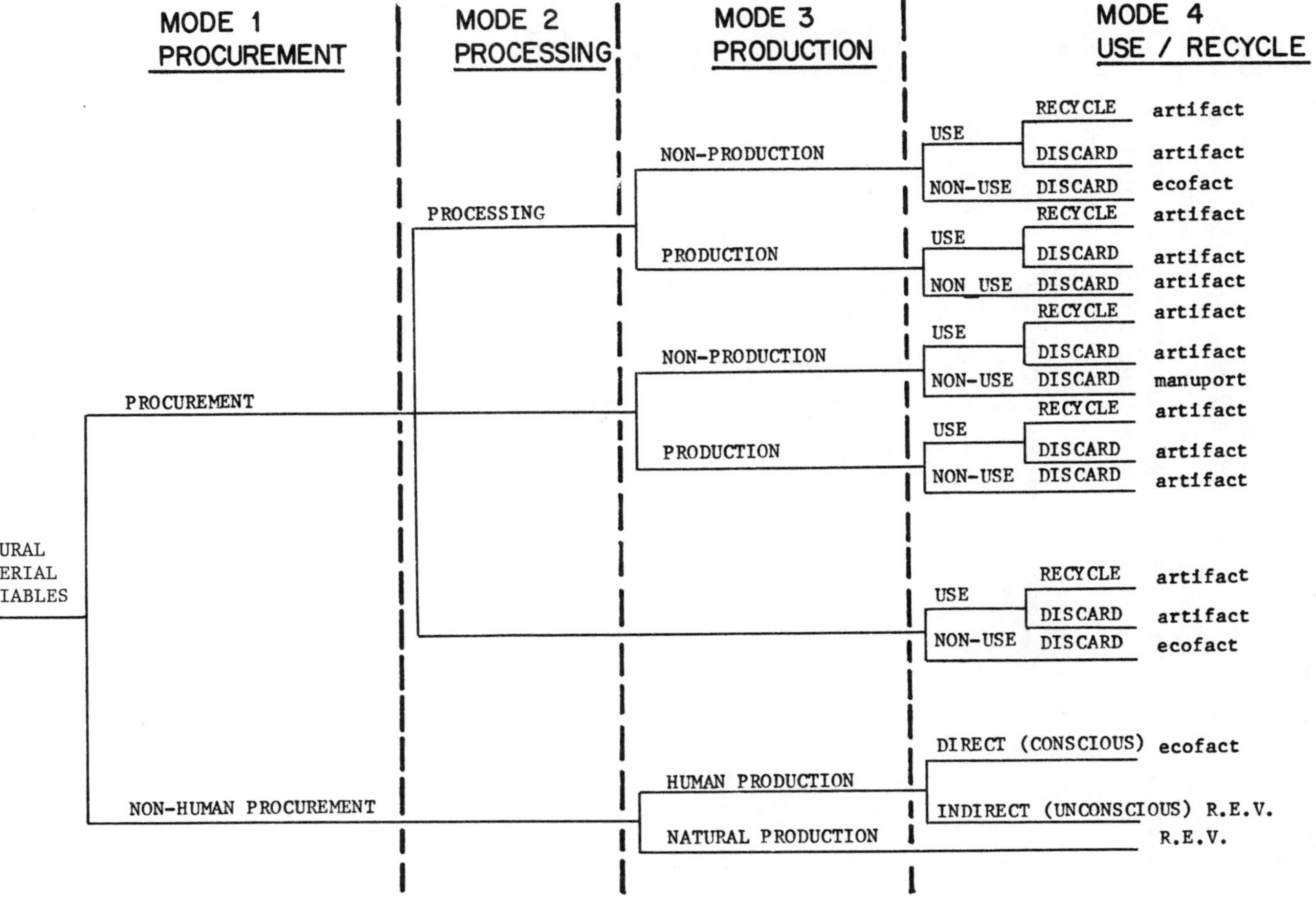

Note: The Recycling mode can feedback to any preceding mode.

R.E.V. – Relevant environmental variable

Fig. 2.1. The sequence of modalities which result in archaeological data

Finally, we are left with data which were not directly collected
(i.e., used by the system) but which, nevertheless, provide information about
the system (i.e., pollen grains blown onto a site during its occupation,
alluvial sediments, arroyo erosion, Carbon-14 ratios, and paleo-climatic
conditions). These kinds of data will tentatively be referred to as *relevant
environmental variables*. They are subject to a totally different set of
causal mechanisms (see Fig. 2.2). Data relevant to the elucidation of most
archaeological problems must include more than "artifactual" and "ecofactual"
information alone. *Relevant environmental variables*, therefore, would include
information which is necessary to problem-solving, but which is not necessarily
cultural or prehistoric in nature (i.e., existing landforms, climate, flora,
and fauna). The importance of *relevant environmental variables* should not
be overlooked, since these data complete a systemic model of cultural-
ecological interchanges.

Natural, Raw Material	Procurement	Processing	Production	Use-Reuse	Descriptive Category	Taxonomic Category
Chert Nodule	X	X	X	X	Core	Artifact
Chert Nodule	X	X		X	Utilized flake	Artifact
Chert Nodule	X				Transported stone	Manuport
Chert Nodule	X	X			Debitage	Ecofact
Chert Nodule	X			X	Hammerstone	Artifact
Clay	X	X	X	X	Pottery	Artifact
Shell (valve)	X	X			Food (by-product)	Ecofact
Shell (valve)	X	X	X	X	Ornament (bead)	Artifact
Wood (log)	X	X			Sawdust-Wood Chips	Ecofact
Wood (log)	X	X	X	X	Canoe	Artifact
Wild Animal--bone	X	X			Disarticulated (by-product)	Ecofact
Wild Grass	X	X--harvest	X	X--food	Corn (domesticated)	Artifact
Wild Bovine	X	X	X	X	Cattle (domesticated)	Artifact
Corn (domestic)	X	X	X	X--food	Tortilla	Artifact

Fig. 2.2. Examples of Relevant Environmental Variables

The rationale for the development of an energy-exchange continuum
model is twofold. First, general systems theory has provided a conceptual
vehicle for understanding the complexity of cultures both past and present.
Ecology has laid open the myriad of pathways through which a culture interacts with its social and natural environment. The evolution of state-level
industrial societies has greatly magnified the significance of

cultural-environmental interchanges. A semantic taxonomy is needed in which
the material variables (archaeological data) can themselves be modeled within
a systemic framework for proper conceptualization. Prior definitions of
"artifacts" and "ecofacts" were either too restrictive or too inclusive to be
useful in this regard.

Second, contemporary archaeologists are striving to transcend "etic"
strategies in order to actualize "cognitive archaeology" (e.g., Schiffer 1976).
An analysis of artifacts and attributes using the framework suggested here
may result in clusters which are more reflective of "emic" categories (i.e.,
a given piece of debitage may or may not be recognized as having a distinct
reality based on its form). This line of inquiry offers some interesting
insights into the distinctive way in which humans modify naturally-occurring
materials in their environment. The question could be posed as to whether
domesticated plants and animals are artifacts. Initially, wild animal species
were *procured* from the environment, diet and living (storage) conditions
reflected *processing*, selective breeding involved the *production* mode, and the
use of the animal was reflected in succeeding generations as well. The results
of domestication both in plants and animals can be empirically determined in
archaeological materials. Humans may, therefore, be the only animals capable
of *consciously* (production mode) manipulating gene pools (*living* artifacts).
At a time when studies in ethnology and human ecology cloud definitions of
"culture" and "human," the manufacture of "living artifacts" may be a talent
unique to human culture.

In summary, it is suggested that the term "artifact" as presently
used is much too nebulous for the rather precise data sets needed for opera-
tionalizing a systemic approach in archaeology. The various modalities men-
tioned above are the direct results of *processes* affecting cultural trajectories
or changes; the recognition of these modalities in the data should facilitate
the study of *culture process*.

3   The Primary Cultural Processes: Towards a Unified Genetic Model of Culture Change

Henry C. Koerper and E. Gary Stickel

INTRODUCTION

"Process" is a term which appears frequently in the terminology of scientific archaeology. In fact, it has been used to describe a so-called "processual school" in contemporary archaeology which is seeking to develop a strictly scientific approach to the study of humankind's past (cf. Watson, LeBlanc, and Redman 1971; Hole and Heizer 1973). However, the term itself (at least with regard to archaeology) has been little discussed, and subsequently its meaning remains unclear.

Varying degrees of utility characterize "process" as this term is encountered in the archaeological literature. David Clarke (1968:42), for instance, defines process as ". . . a vector which describes the series of states of an entity or system undergoing continuous change in space or time." Still other meanings of "process" are presented by Hole and Heizer:

> One can readily see that process means two quite different things. First, it may refer to a *sequence of events*. Second, it may refer to the *causes* of the sequence of events. In both meanings, process is conceptually linked with the states or conditions of the things under observation at different times. As process is used in archaeology (cf. processual analysis, process archaeologists), it refers to an analysis of the factors that cause changes in state; that is, from "barbarism" to "civilization" [1973:439; emphasis ours].

We do not conceive of a process as a "vector" which *describes* the series of states of a changing system as suggested by Clarke. Neither do we conceive of a process as simply a *sequence* of events or an *analysis* of the causes of a sequence of events as suggested by Hole and Heizer. Process or processes ought to refer to the *factors* that cause changes in states of cultures across *time*.

Different levels of processual analysis can be recognized. For instance, at a fundamental level there is cultural change as determined by the "primary processes," and at the most general level there are the lawful

---

[1]We gratefully appreciate reviews and criticisms of this paper by Professors Joseph B. Birdsell, University of California, Los Angeles, and Irving Taylor, University of California, Riverside.

principles governing the operation of culture in general (e.g., cultural
growth, cultural equilibrium, or cultural devolution).

Our subject here is the *primary processes* of culture change. These
basic processes will be viewed as analogous to the mechanisms of biological
(phylogenetic) evolution: (1) mutation, (2) gene flow, (3) sampling
phenomena such as genetic drift, and (4) natural selection. Such a strategy
will allow us to evaluate the several "primary processes" as they are variously dealt with in the literature. It will also facilitate an understanding
of how best to conceptualize those primary processes.

We suggest that the basic causes of culture change are the *primary
processes*. These, for David Clarke, are "inevitable variation, multilinear
development, invention, diffusion and cultural selection" (Clarke 1968:22).
Both Hole and Heizer (1973:81-83, 439) and Fagan (1975:312) echo David
Clarke's primary processes, although Fagan has dropped "multilinear development" as a process.

Generally, it is our view that the primary processes have been
either neglected or improperly treated in the literature. In the hope of
clarifying the issue, the mechanisms of cultural evolution (primary processes)
are viewed here as paralleling the mechanisms of biological evolution. Such
a biological analogy has been both discounted and commended as a useful
analytic device by anthropologists:

> The analogy between biological and sociocultural evolution breaks
> down at a number of crucial points. First, the "traits" or "units"
> that evolve in bioevolution--namely genes and organisms--replicate
> themselves in a fashion quite unlike the replication of cultural
> patterns. Biological reproduction consists of a formation of
> daughter genes which govern the growth of new organisms. Socio-
> cultural systems do not contain genes and hence they do not undergo
> anything similar to biological reproduction [Harris 1971:150].

By *analogy*, Harris seems to imply a need for the existence of direct
functional correspondences. We use the term in its traditional sense to suggest that tacit similarities exist between things that imply further correspondences. Mark Leone recognizes that cultural analogues to the biological mechanisms of evolution have heretofore been incomplete, but he also
wisely suggests that such analogues are potentially useful (Leone 1972:26).

One might intuitively sense at least a limited value in outlining
these analogues, for parallels between cultural and biological evolution
begin with cultures and breeding populations; both are open dynamic *systems*
which tend toward growth and differentiation. (Open systems are ones which
exchange matter, energy, and information, in any form, with their environments.) In the course of macro-evolution, cultures and breeding
populations have generally become more efficient energy-capturing systems,
and in their evolution (which is characterized by increasing heterogeneity
and complexity of structure) they must both operate counter to the second
law of thermodynamics.[2]

---

[2]The second law of thermodynamics states that in closed systems (in
which there is neither import nor export of energies), the general course of

Further parallels between biological and cultural evolution involve the following two points: (1) the long-run evolution of a breeding population generally involves the accumulation of progressively more complex information in the form of genetic codes; and (2) as cultures evolve (i.e., as they generally become more efficient energy-capturing systems), the ideas (information) they carry become greater in number and increasingly more complex. Just as biological evolution is characterized by changes in gene frequencies in breeding populations through time (across generations), so cultural evolution may be critically viewed as changes in idea frequencies in cultures through time.

Cultures will differ from one another in the degree to which they exhibit varying frequencies in ideas. A culture which has changed through time is a culture whose idea frequencies have altered. Ideas--not unlike genes--are bits of information. Ideas are cultural codes that prescribe such things as how to knap a particular projectile point or how a male in a matrilineal society behaves vis-à-vis his mother's brother.

In biological evolution, the processes of mutation, gene flow, and sampling phenomena such as drift operate directly to alter gene frequencies, while the selective process operates on phenotypes (the behavior and physical traits of organisms) to indirectly change gene frequencies within breeding populations.[3] Similarly, in our genetic analogue model of cultural change, cultural mutation, cultural flow, and cultural drift (sampling phenomena) may operate directly on the idea frequencies in cultures. Cultural selection operates on "cultural phenotypes" (i.e., the *means* of system operation commonly called human behavior and all *media* such as artifacts utilized by a system) to indirectly change idea frequencies. The functional relationship between ideas and their manifestations (cultural phenotypes) can allow prehistorians to infer cultural evolution from observable changes in frequencies of such things as pottery types, dwelling foundations, and settlement patterns in the archaeological record.

THE PRIMARY CULTURAL PROCESSES

The following analogues are proposed between biological and cultural evolutionary processes:
(1) Genetic mutation and cultural mutation: Cultural mutation involves discovery, invention, and stimulus diffusion.[4]

---

physical events is one in which a certain quantity, *entropy* (which may be taken as the measure of unavailable energy in a thermodynamic system), increases to a maximum, and the process eventually stops at a state of equilibrium. This involves the leveling-down of differences in a closed system to a state of maximum disorder, and hence entropy might be thought of as a measure of disorder. Parenthetically, negative entropy is characteristic of open systems. The universe is presumably a closed system; therefore, physical nature should theoretically gravitate toward disorder and homogeneity. The final result would be a universe with evenly distributed matter at the same low temperature or energy state. This is the so-called "heat death of the universe."

[3]For a clear explanation of the biological mechanisms of evolution see Birdsell 1975.

[4]Stimulus diffusion refers to the generation of new ideas and

(2) Gene flow and cultural flow: Cultural flow refers to (or involves) what has been traditionally described as diffusion, migration, and acculturation.

(3) Genetic drift and cultural drift: Cultural drift refers to variation as a result of sampling error phenomena.

(4) Natural selection and cultural selection: Cultural selection involves selection for or against ideas via their manifestations (systems' means and media).

Cultural Mutation

Cultural mutation may be defined as the creation of new ideas (information) within a culture as a consequence of discovery, invention, or stimulus diffusion. Biological mutation and cultural mutation are born of unlike circumstances, but the important point to note is that they represent the initial source of biological and cultural deviation, respectively. In other words, they result in new information, which must then be subject to selective pressure, flow, or drift (or any combination of these) if ever the mutant gene or new idea is to be of evolutionary significance.

Cultural mutation takes place within human minds (the ideational level) and may be expressed as new behavior (e.g., an increased respect for matriarchal authority or a new type of loom). Whether a new idea increases in frequency or not usually depends upon the cultural/ecological factors which select for or against the expression of the idea.

Mutations may or may not be adaptive in a given situation confronting a culture. Their presence and use, however, may well determine (given a specific situation) the survivability of a culture. Hence, cultural mutations are important in the evolution of culture. The rapidly increasing complexity of modern culture is the direct result of an acceleration of this process through experiment and research.

Cultural Flow

Cultural flow is the process which causes alterations in idea frequencies as a result of the transference of ideas from one culture to another. These idea transferences are brought about by simple diffusion (i.e., the movement of ideas from one culture to another without population movements), the actual migration of people into another receiving culture, acculturation, etc.

There have been previous attempts to develop a cultural analogue to gene flow; for example, "cultural flow" has been defined as ". . . intra-generational and intergenerational transmissions of ideas, modes of behavior, and the material products of behavior" (Dunn 1970:1042). But it is inappropriate to lump the basic units of culture (ideas) together with the cultural manifestations of ideas (the behavioral and material "phenotypes") in transmission, and (more importantly) Dunn's transmission is more an analogue of

---

their possible cultural manifestations (e.g., behavior and artifacts) by members of one culture who have been inspired by observing similar but *not* identical ideas in other cultures (see Kroeber 1940).

biological sexual recombination than anything else. Schneider (1977:11, 18)
has compounded the confusion by equating his "standardization" or "socialization" with Dunn's "cultural flow." The biological counterpart, according to
Schneider, is "inbreeding," which is described as a "process" (Schneider 1977:
11). Such a notion fails to advance our understanding of cultural process.
Inbreeding, by itself, does not change gene frequencies, and "socialization,"
while it is responsible for the transmission of ideas, does not by itself
alter the frequencies of these cultural units. Hence, inbreeding is not a
mechanism of biological evolution, and socialization is not a primary process of cultural evolution.

<u>Cultural Drift</u>

        Cultural drift leads to changes in idea frequencies due to sampling
error. As conceived here, cultural drift is analogous to both genetic drift
and the founder effect. It was first proposed archaeologically by Binford
(1963), who suggested that the variation in stylistic elements in prehistoric
Michigan from sub-region to sub-region was the result of notions of design
forms drifting away from a once-coherent style pattern, due to an increasing
lack of intercommunication over time between populations who no longer exchanged ideas about design execution.

        Additional examples of unsuccessful efforts by archaeologists to
develop a cultural analogue to genetic drift can be cited:

> Inevitable variation is somewhat similar to the well-known phe-
> nomenon of genetic drift in biology. As people learn the be-
> havior patterns of their society, inevitably some minor dif-
> ferences in learned behavior will appear from generation to
> generation, which, minor in themselves, accumulate over a long
> time, especially if the populations are rather isolated [Fagan
> 1975:312].

> The population that moves into a new environment with a new set
> of techniques develops further along lines laid out by . . .
> [an] initial invention. The invention is the "kick" in the
> system, which is amplified by continued use and elaboration.
> Metaphorically and physically the population drifts farther
> away from the parent group [Hole and Heizer 1973:464].

In the first quote, the "minor differences" are not properly ascribed to
cultural mutation, and Fagan does not acknowledge the random nature of drift.
Although Hole and Heizer correctly link "random change" with genetic drift
elsewhere (Hole and Heizer 1973:81), here "drift" is the result of cultural
mutation ("initial invention") and cultural selection ("continued use and
elaboration") and should not be viewed as a random process.

        Charles Cleland (1972) has used the term "style drift" (which he
calls a "process") to discuss a sixty-year period of change (rearrangements,
additions, and/or deletions) in discrete design elements in French Jesuit
finger-ring motifs. He attributes the observed changes to the "passive
process" of "style drift," coupled with "positive selection" (which involves
an "active" cultural choice). The illustrations of the various changes
leave no doubt that a "stylistic breakdown or a reduction of structures"
occurred over time. But the term "style drift" actually describes the

effects of cultural mutations within a context of relaxed selective pressures, and as such recalls biology's primary mutation effect. Again, drift
has been linked to non-random combinations of processes.

        For small human populations, it is reasonable to suggest that
differential reproductive rates of families (which are the primary encultura-
tive institutions) might initiate shifts in idea frequencies. This would
be especially true in cases where ideas associated with certain cultural
characteristics varied between families. For such ideas, which would be
unrelated or only remotely related to the selective factors contributing to
a specific family's greater reproductive success, the shifting frequencies
might be seen to drift due to chance.

        Cultural drift in the above example is an analogue of "genetic
drift" proper, which is subsumed under the rubric of sampling error phenomena
along with the "founder effect" (also referred to as the "bottleneck effect").
The founder effect, unlike genetic drift proper, is included in Sewall
Wright's (1955) category I3, the category of random unique events. A cultural analogue of the founder effect is presented below.

        It is reasonable to suggest that when a small population segment
separates spatially from its parent human system, that segment might not be
culturally representative of the larger parent population, especially in
terms of at least some of its idea frequencies. Consequently, another
sampling phenomenon, akin to biology's "founder effect" or "bottleneck effect,"
could be said to have occurred. Dunn (1970:1042) was close to a correct
conceptualization of the cultural founder effect in his example of colonists
who brought only some fraction of the total cultural variation available to
them from a parent population to their new settlement, but he did not conceive of this process in terms of changing ideas or information frequencies.

        Schneider (1977:18), in an attempt to follow Dunn, offered a morbid
example of a cultural founder effect. An atomic war occurs in which all
citizens of the United States are killed save for a small enclave of illit-
erate sharecroppers; the subsequent culture then lacks writing. Although
this hypothetical situation differs from Dunn's example in that a disaster
has caused this unique reduction of population numbers, a model of fluctuat-
ing idea frequencies, which would add clarity and insight to the examples,
is lacking in both cases.

        In our estimation of the four primary processes, cultural drift
generally has the least impact on culture change. Nevertheless, it is a
source of some of the observable cultural variability in past and present
human systems.

Cultural Selection

        The very fact that there is an archaeological record composed of
a vast number of remains of prehistoric cultures attests to the fact that
there were differences in the ability of these cultures to survive. This
differential survivability is largely due to *cultural selection*. Cultural
selection involves changes in idea frequencies due to selection for or against
cultural "phenotypes" (i.e., the cultural *means* of system operation, also
called human behavior, and cultural *media* such as artifacts). (The term

cultural selection should not be confused simply with attitudes, preferences, and personal choices, but should refer to that general selective process which results in changes in idea frequencies in culture.)

There are two major points to be made in discussing selection. First, the manifestations of ideas (cultural phenotypes, or traits), in a milieu of inter- and intra-cultural competition, are adaptive, neutral, or maladaptive in varying degrees so far as the survival of a culture is concerned. Selective coefficients for or against a trait will have corresponding effects upon the idea of which the trait is an expression. The parallel with natural selection is readily seen, for selective factors operate on biological phenotypes and, in selecting for or against the phenotypes, alter the gene frequencies of the alleles which partially determine the phenotypes. Second, just as in human biological evolution, the "environments" of human cultural evolution in which selective factors are operative are *total systems* which include both natural and other human systems components.

In analyzing systems literature in archaeology, it is apparent from the numerous references to "adaptation" that those employing a systems model either knowingly or unknowingly place their emphasis upon the "primary process" of cultural *selection*. Systems theory, as it is used in archaeology, may accordingly be viewed largely as directed at positing the workings of important relationships which help to account for the selective factors in the adaptive or even maladaptive orientations of cultures. These points may be succinctly demonstrated by reference to a paper by Lewis Binford.

In "Post-Pleistocene Adaptations," Binford (1968b) argues persuasively that the cultural transition from the Paleolithic to Mesolithic to Neolithic is not to be understood as simply the result of a direct stimulus response to environmental change. Binford uses a systems approach to generate testable hypotheses in which demographic variables are given causal status to explain certain post-Pleistocene adaptations, especially those which resulted in the Neolithic, with its development of agriculture and animal husbandry.

Binford's model develops in the following way. First, the equilibria systems of hunting and gathering cultures regulate their population densities to below the carrying capacity of their environment. Thus, there are no compelling food stress factors which necessarily force hunting and gathering people to increase their food supply. The crucial question revolves around the conditions under which an increase in food supply conveys an adaptive advantage. Specifically, in the view of the present paper, the question revolves around the factors which determine a shift in selection coefficients. These shifts in selection coefficients resulted in an increased frequency of cultural information regarding more proficient food procurement (e.g., agriculture).

Binford describes two situations. In the first, two different kinds of cultures would occupy contiguous geographic zones. If the exploitative strategy of the dominant culture could be implemented in the zone of the other culture(s), then encroachment on the weaker residents would result. A consequence of this occupation would be an increased population

density in the zone. Such population density might then lead to an adaptation involving more efficient food production and/or an increased regulation
of birth rate.

In Binford's second situation, the subsistence strategy of a
dominant culture (with a rapidly growing population) would not be trans-
latable into the adjacent culture's less populous zone. Binford suggests
that differential sedentism would be the cause of the relevant differential
population patterns. He proposes that the relatively sedentary nature of
the dominant culture would reduce the need for wide-ranging mobility,
especially for women. The result of this decreased mobility would be a
diminished selective advantage for cultural means to control population
growth (such as infanticide). Very simply, sedentism would allow women to
have more offspring than they could transport and maintain with a more mobile
system. The dominant culture's overpopulation would lead to encroachment
into the second zone. This intrusion would do violence to the already
existing human density equilibrium established in the zone, resulting in
diminishing food resources for the less populous culture. Within such a
context of burgeoning population, strong selective pressures for increasingly
efficient exploitative techniques would affect both groups, particularly at
the population frontier or "tension zone." Binford goes on to hypothesize
that the initial stages of agriculture should not be sought in the center of
the natural habitat zones of potential domesticates (as formerly thought),
but at such intersections as just described.

Verification of the foregoing model is another issue (Wright 1971);
our point here is that these hypotheses are clearly set within a systems
frame, and the role of selection is given explicit emphasis. Moreover,
Binford emphasizes the interplay between the social environmental and the
natural environmental components of the total system (i.e., the total environment within which selection takes place). Thus, this model facilitates the
study of the effects of cultural selection on the evolution of culture.

THE TOTAL ENVIRONMENT

Willey and Sabloff (1974:189-190) have noted that Binford, in his
systemic view of culture, lacked (at least in 1962) a systematic, holistic
view of cultural systems in relation to natural environments. However,
Binford's regard for the significant role played by the natural environment
is evident in his statement that Steward's cultural ecology (the compara-
tive study of cultures with variable technologies in differing environments)
is ". . . the most valuable way of increasing our understanding of processual
interpretation" (Binford 1962:217). It is quite understandable that Binford's approach has led to an emphasis on a "cultural ecological approach"
in archaeology.

With the natural environment given its proper role by other
scholars (see especially Clarke 1968 and Flannery 1968), clarity is brought
to the issue of what ought to constitute the total "system" in archaeological systems theory. From this insight follows our view that systems
theory can function effectively to allow an analyst to construct a total
environmental model. This total environment is comprised of both

relevant[5] cultural (all important interacting human systems) and natural
environmental subsystems. It is within this *total environmental system* that
selective factors may be understood.

<u>PROCESSES AND THEIR RELATIONS</u>
<u>TO SYSTEMS THEORY</u>

The relevance of primary processes to a systems approach, which
aims at explanation, will be explored here. David Clarke (1968) considers
systems theory to be the vehicle which relates processes to culture change--
a position with which we concur. In our view, systems theory gives primacy
to the process of selection. This is an important strategy, for selection
is generally the most important of the processes which cause culture change,
once the initial variation is introduced. Consequently, this view lends
itself to relatively sophisticated approaches to the understanding of culture change, especially where change is largely endogenous (i.e., where
change is not due to a form of cultural flow; if change were exogenous, then
it would involve cultural flow). However, as important as selection may be,
the other primary processes cannot be ignored if an understanding of cultural
evolution is to be achieved. Overenthusiastic proponents of systems theory
often impart to their students the false impression that other "primary
processes" might best be left out of the larger picture. Flannery, for
example, in his "cybernetics approach" to the transition from food-collecting
to horticulture in Mesoamerica, praises a systems approach which, he argues,
obviates any need to attribute cultural evolution to "discoveries," "inventions," "experiments," or "genius" (Flannery 1968:85). The systems approach
advocated here takes these potentially important phenomena into account.

"Process" archaeologists have also reacted against diffusionist
models of culture change. Some traditionalists may certainly be legitimately
criticized for the ease with which they invoke diffusion in specific cases
to account for similarities between widespread cultures. But diffusion, as
a kind of "culture flow," should not be overlooked as a processual vector
in numerous cases of systemic culture change--a reading of Linton's classic
paper "One Hundred Per Cent American" ought to bring even the most reactionary
of anti-diffusionists back to reality (Linton 1937).

The interplay of information, primary processes (especially selection), and the total system forms the basis for any systems approach we would
advocate. We also suggest that the primary cultural processes operate in
conjunction with the primary biological processes (gene mutation, gene flow,
genetic drift, and natural selection) to cause *human systems* in their
entirety to change. As viewed here, human systems are systems involving both
biological and cultural variables. The biological variables are those which
comprise human organisms only. The biological processes operate to both
cause and affect genetic variability within the biological component of any
given human system (the system's human population), just as cultural processes affect variability within all of the cultural components of any given

---

[5]We equate the "relevant environment" with Binford's ". . .
effective environment which designates those parts of the total environment
which are in regular or cyclical articulation with the unit under study"
(Binford 1968b:323).

human system. Thus, these two sets of processes cause human systems to change (develop, survive, or perish) in their *entirety* (Fig. 3.1).

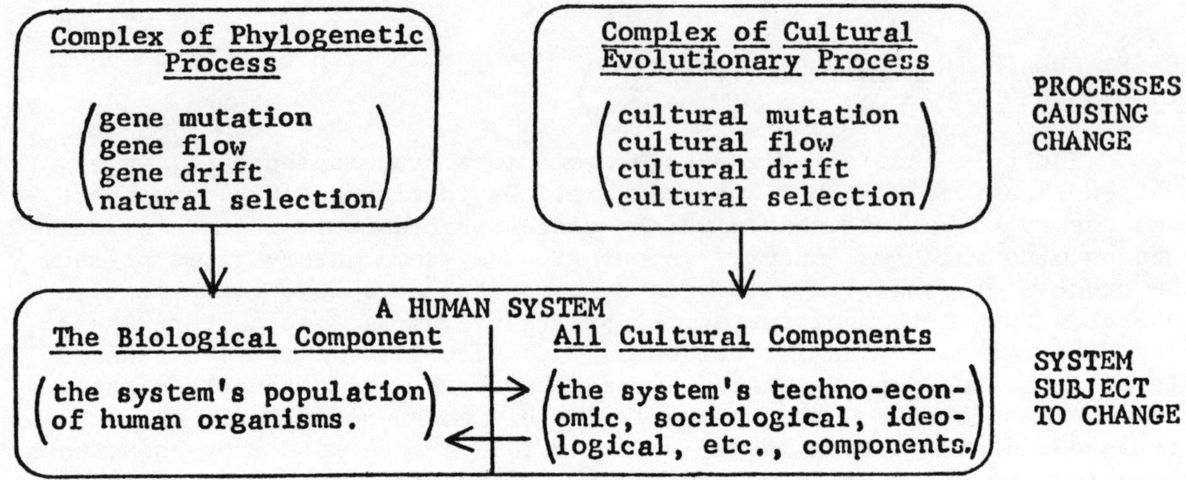

Fig. 3.1. *The combined effects of phylogenetic process and cultural process produce total human system change.*

<u>PROCESS AND EXPLANATION</u>

We must now ask the following question: Does the identification of the "primary processes" explain cultural change? We have already compared the primary processes of cultural change to the mechanisms of phylogenetic change. But the biological mechanisms are not really explanatory in nature. A simple statement that a particular organism evolved because it was subject to the mechanisms of phylogenetic evolution bypasses both the scientific testing of particular explanations of biological change and the subsuming of particular instances under discoverable lawful principles. In science there is a close association between "process" and "explanation"; however, the identification of primary processes cannot be equated with explanation. Steward, for example, has noted that

> Childe's transfer of the Darwinian formula to cultural evolution
> will not evoke challenge. Variation is seen as invention, heredity
> as learning and diffusion, adaptation and selection as cultural
> adaptation and choice (Childe 1951:75-79). It must be stressed,
> however, that all universal laws thus far postulated are con-
> cerned with the fact that culture changes--that any culture changes--
> and thus cannot explain particular features of particular cul-
> tures. In this respect, the "laws" of cultural and biological
> evolution are similar. Variation, heredity and natural selection
> cannot explain a single lifeform, for they do not deal with the
> characteristics of particular species and do not take into
> account the incalculable number of particular circumstances and
> factors that cause biological differentiation in each species
> [Steward 1955:18].

We maintain that the study of primary processes will provide a greater understanding of cultural change, and will thus facilitate explanation.

<u>SUMMARY AND CONCLUSIONS</u>

This discussion of the primary processes of cultural change has hopefully contributed to archaeological perspectives in several ways. We have emphasized the fact that there has not been a clear treatment of primary processes in the literature, and have tried to demonstrate the feasibility of constructing cultural analogues to biological evolution. The macroevolutionary histories of both cultures and breeding populations are played against a backdrop of changing total environmental conditions and a universe which is running down both structurally and dynamically. Consequently, a struggle for free available energy ensues. The result of such a struggle is the emergence of progressively more efficient energy-capturing systems--that is, systems which are generally better adapted to an increasingly competitive world.

The forms of both cultural manifestations and biological organisms are largely determined by coded information. Analogous mechanisms of change in cultural and biological evolution alter the frequencies of various types of information through time as a response to the requirement that both cultures and breeding populations accumulate negative entropy. The primary cultural mechanisms or processes are identified as cultural mutation, cultural flow, cultural drift, and cultural selection. We also emphasize the fact that the proposed model of primary processes is not in itself truly explanatory. However, the model does provide some necessary clarification and illumination along the road toward explanation. We believe that systems theory is useful for studying the shifting selection coefficients which play a significant role in cultural change. Therefore, systems theory should aid in cultural historical explanation and in the testing of propositions for the verification of lawful principles of cultural change.

4   Energy and Symbolism in Mortuary Practices

Joseph A. Tainter

INTRODUCTION

The study of mortuary practices (as in so much of social science) has been characterized by some diversity in how anthropologists have interpreted mortuary behavior in general. There has been a contrast between interpretations which have stressed the psychological aspects of the ritual, such as beliefs in an afterworld, fear of the corpse, and the like (e.g., James 1928), and approaches that have focused on integrating the ritual into a cultural context (Huntington and Metcalf 1979). A basic dichotomy can also be observed within this latter orientation. On the one hand, anthropologists such as Kroeber (1927), Ucko (1969), and Huntington and Metcalf (1979) have stressed the culture-specific nature of mortuary ritual, and emphasized the seemingly idiosyncratic variations in mortuary practice between one group and another. Another group of investigators have argued that the form of a ritual is not an idiosyncratic occurrence, but instead is linked directly to such factors as the age, sex, and status of the deceased (e.g., Wedgwood 1927; Bendann 1930; Binford 1971).

These diametrically opposite views involve more than simply questions concerning the nature of mortuary practices; they also have broader implications related to human symbolism and the potential for sociocultural analysis in archaeology. If mortuary behavior, or any kind of symbolic behavior, can be shown to vary idiosyncratically, dependent entirely upon such factors as emotional whimsy and historical accident, then it can be asserted that the process of symbolizing is not part of a linked, interdependent system, and symbolizing is beyond the realm of scientific explanation. Indeed, the process of symbolizing is so basic to cultural phenomena that a demonstration of idiosyncracy in the process would virtually isolate anthropology from the world of objective, generalizing science. (Unfortunately, it seems that more than a few anthropologists would not be dismayed by such a prospect.) In archaeology, a failure to establish a systemic linkage between symbolism and material phenomena would render mortuary ritual (and indeed most past sociocultural phenomena) undecipherable. I do not mean to suggest that mortuary ritual is in some manner more crucial than other forms of symbolic expression, but I do contend that it is an interesting example with important implications.

<u>SYMBOLISM IN MORTUARY RITUAL</u>

Kroeber, in an influential article (1927), argued that mortuary ritual is an idiosyncratic event. He noted that mortuary customs in aboriginal California were distributed in a seemingly irregular manner, and were not in conformity with other cultural features. He then argued that mortuary practices, along with some other aspects of behavior, display a tendency toward instability because of their lack of integration with other cultural phenomena. He suggested that

> . . . disposal of the dead has little connection with that part of behavior which related to the biological or primary social necessities, with those activities which are a frequent or constant portion of living and therefore tend to become interadapted and dependent one on the other. On the other hand, disposal of the dead also does not lend itself to any great degree of integration with domains of behavior which are susceptible of formalization and codification, like law, much of religion, and social organization [1927:314].

In other words (although he would not have conceived of it in precisely these terms), Kroeber explicitly argued that mortuary symbolism did not form part of a system, but instead should be viewed as "partaking of the nature of fashion" (1927:314). He went on to state that this characteristic of mortuary practices tended to ". . . discredit certain interpretations based upon them" (1927:315).

The data which led Kroeber to this conclusion are, indeed, puzzling. The most interesting of them involve patterns of antagonistic opposition, in which adjacent groups employ the same symbolic form to signify different social personalities. For example, in regard to sub-Saharan Africa, Kroeber made the following observation:

> Thus, river burial is sometimes reserved for chiefs, sometimes for the drowned, sometimes is the normal practice of a group. Tree and platform burial is in certain populations restricted respectively to musicians, magicians, the bewitched, the lightening struck, criminals, and kings; cremation is generally reserved for criminals, but also occurs as the usual practice; exposure is variously in usage, according to tribe, for the corpses of criminals, slaves, children, the common people, the entire population [1927:313].

Kroeber asserted that these examples ". . . constitute a powerful argument for instability" (1927:313).

Ucko (1969), struck by the same kinds of idiosyncracies that Kroeber had noted (but with less interest in their possible theoretical significance) marshalled an impressive array of ethnographic examples designed to show that many of the assumptions that archaeologists have made concerning the meaning of mortuary remains have been questionable. Thus, the traditional archaeological interpretation that elaborate mortuary ritual signifies a belief in an afterlife, and that rich graves with elaborate goods indicate rank differentiation, are shown to be debatable in ethnographic contexts. The intent of Ucko's essay is to impress archaeogists with the lack of systematic linkages between mortuary symbols and their referents. It might be noted that Ucko's cautionary observations concerning the more simplistic types of archaeological interpretations came at a time when the study of prehistoric mortuary remains was shifting in emphasis away from these kinds of inferences (e.g., Stickel 1968; King 1969).

The most recent contribution in this vein is a book by Huntington and Metcalf (1979). These authors, who adopt a highly particularistic approach, acknowledge the existence of what appear to be near-universals in mortuary behavior (such as color symbolism, hair-cutting, noisiness, and symbols of decay), but assert that these universals may arise from disparate cultural matrixes. They argue for a case-by-case analysis as the key to understanding mortuary customs, and believe that "placed in context of a particular ideological, social, and economic system, rituals of death begin to make sense in a way that they cannot if we pursue elusive cultural universals" (Huntington and Metcalf 1979:58).

The studies cited above, while displaying some variety in concept and purpose, are nevertheless characterized by a basic similarity in their underlying approach. At their core is a particularistic, culture-specific framework for mortuary analysis. To Kroeber, Ucko, Huntington, and Metcalf, the outstanding characteristic of mortuary rituals is their culture-by-culture variability; divorced from the cultural context which produced them, they are supposedly inexplicable. Kroeber carried this particularistic perspective to what seems an extreme when he suggested that mortuary rites are inscrutable even *within* their cultural contexts. The only conclusion that one can draw from this suggestion is that a cross-cultural theory of mortuary practices is impossible, and mortuary behavior lies beyond the realm of scientific analysis; only by subjective, culture-specific inference is it possible to appreciate why a particular people dispose of their dead the way they do.

It is curious that those ethnologists who have espoused a more objective approach in anthropology have been silent about attempts to remove mortuary practices from the realm of valid generalization. Perhaps there are more important battles to be fought in ethnology. But archaeologists, who frequently contend with human burials as a major part of their data base, have not overlooked the challenge. The potential for sociocultural analysis provided by burials is unsurpassed; Christopher Peebles was not exaggerating when he wrote that "a human burial contains more anthropological information per cubic meter of deposit than any other type of archaeological feature" (1977:124; see also Tainter 1978:110). As long as mortuary behavior is seen as unrelated to other cultural phenomena (Kroeber 1927), essentially inscrutable (Ucko 1969), or highly particularistic (Huntington and Metcalf 1979), archaeology's potential for contributing to the study of social organization is diminished. The loss is not just to archaeology, but to anthropology as a whole.

It should come as no surprise, therefore, that the major challenge to the particularistic approach has been mounted by archaeologists. In 1962, Lewis Binford published a now-classic paper entitled "Archaeology as Anthropology," in which he argued (among other things) that the complexity of material indicators of social rank, as well as their association with individuals and groups, will vary according to the complexity of the society. E. Gary Stickel (1968) later expanded upon this framework and applied it

directly to mortuary research. Stickel argued that grave associations will
be patterned differently in egalitarian and ranked societies. In an
egalitarian society, Stickel (1968:217) proposed, grave associations should
reflect the following characteristics: (1) a predominance of artifacts that
served primarily in the technological sphere; (2) ranking symbols of the
sort attainable by individual achievement; (3) grave associations that dis-
tribute differentially among individuals rather than groups within such
ranked groups as exist; and (4) the non-inheritance of ranking symbols,
potentially signified by the inclusion of such items in graves. However,
in a society characterized by class ranking, grave associations should
pattern in a contrasting fashion; there should be (1) an increased frequency
of ranking symbols; (2) a possession of specific ranking symbols by groups;
(3) considerable individual variation in mortuary associations; and (4) an
inheritance of ranking symbols. In addition to its intended application to
mortuary remains, Stickel's formulation suggests that there may be dimensions
of mortuary symbolism which transcend the culture-specific idiosyncracies
emphasized by Huntington and Metcalf; perhaps "elusive cultural universals"
(Huntington and Metcalf 1979:58) are worth pursuing after all.

After the appearance of Stickel's article, Arthur Saxe completed
a doctoral dissertation (1970) in which he developed a series of interlinked
hypotheses concerning mortuary symbolism. These hypotheses are universal in
their scope; i.e., they are applicable beyond the limitations of specific
cultural contexts. Saxe, utilizing anthropological role theory (Goodenough
1965) as well as techniques of componential analysis, proposed (among other
things) that individuals of low social significance will generally be accorded
a mortuary treatment involving fewer positive symbols than more significant
members of the same society (1970:69). By using componential analysis and
focusing on the number of contrasting mortuary symbols used by a society to
signify persons of different rank, Saxe was able to propose two measures of
sociocultural complexity (1970:75, 112). Saxe tested his hypotheses against
a limited ethnographic sample of three cases; some received support, while
others remained unconfirmed. One hypothesis that did receive support in this
sample of three cases was the following:

> To the Degree that Corporate Group Rights to Use and or Control
> Crucial but Restricted Resources are Attained and/or Legitimized
> by Means of Lineal Descent from the Dead (i.e., Lineal Ties to
> Ancestors), Such Groups Will Maintain Formal Disposal Areas for
> the Exclusive Disposal of Their Dead, and Conversely [Saxe 1970:
> 119].

Lynne Goldstein (1976) subsequently tested this idea on a more extensive
sample of thirty cases and found consistently positive results.

In a 1971 cross-cultural survey of mortuary practices, Binford
set out to test the following two propositions:

> There should be a high degree of isomorphism between (a) the
> complexity of the status structure in a socio-cultural system
> and (b) the complexity of mortuary ceremonialism as regards
> differentiated treatment of persons occupying different status
> positions [and] there should be a strong correspondence between
> the nature of the dimensional characteristics serving as the
> basis for differential mortuary treatment and the expected cri-
> teria employed for status differentiation among societies arranged
> on a scale from simple to complex [1971:18-19].

To rephrase the latter proposition, Binford argued that age and sex should
commonly serve as bases for mortuary distinctions among egalitarian hunters
and gatherers, while differential social position should more frequently be
the basis for such distinctions in more complex societies.

Binford's ethnographic sample (as he acknowledged) was not ideal.
Since it was not possible to directly measure social complexity from the
ethnographic literature, it was done indirectly by noting means of subsistence;
Binford followed this procedure because of the "generally accepted correla-
tion between forms of subsistence production and societal complexity" (1971:
18). Subsistence modes were grouped into four categories: hunters and
gatherers, shifting agriculturalists, settled agriculturalists, and pastoral-
ists.

Despite the problem of sampling, the results obtained were significant. Binford considered the following points to be demonstrated:

> (1) The specific dimensions of the social persona commonly given
> recognition in differentiated mortuary ritual vary significantly
> with the organizational complexity of the society as measured by
> different forms of subsistence practice.

> (2) The number of dimensions of the social persona commonly given
> recognition in mortuary rituals varies significantly with the
> organizational complexity of the society, as measured by dif-
> ferent forms of subsistence practice.

> (3) The forms, which differentiations in mortuary ritual take,
> vary significantly with the dimensions of the social persona
> symbolized [1971:23].

Binford concluded his discussion with the following summary statement:

> These findings permit the generalization that the form and structure
> which characterize the mortuary practices of any society are con-
> ditioned by the form and complexity of the organizational char-
> acteristics of the society itself [1971:23].

Binford also noted that his findings refuted Kroeber's assertion that forms
of burial are not systemically integrated with such other cultural phenomena
as subsistence practices or social features (1971:15).

In the same paper, Binford also observed that the form of a mortuary ritual will be partially determined by the size and composition of
the social aggregate recognizing obligatory status responsibilities to the
deceased. He proposed that such a larger array of duty-status relationships
(characteristic of persons of higher rank) would entitle the deceased to
greater corporate involvement in the act of interment, and to a higher
degree of disruption of normal community activities for the mortuary ritual
(1971:17, 21). I have since suggested that both the degree of corporate
involvement and the degree of activity disruption should be correlated with
the amount of energy (or labor) expended in the mortuary act (Tainter 1973);
thus, higher social rank will correspond to greater amounts of corporate
involvement and activity disruption, and this should result in the expenditure of greater amounts of energy in the interment ritual. Energy expenditure should, in turn, be reflected in such features as the size and elaborateness of the interment facility, methods of handling and disposing of the

corpse, and the nature of grave associations. This proposition has been
tested on an extensive ethnographic sample, and in a set of 103 cases, the
energy-expenditure argument was not contradicted once (Tainter 1975).

Therefore, there are compelling arguments and consistent ethnographic support for questioning the extent to which the study of mortuary
practices must be particularistic in its approach. The notion that burial
procedures can only be understood in a particular cultural context no longer
seems tenable. Those "cultural universals" now appear much less elusive.

ENERGY, SYMBOLISM, AND COMPLEXITY

The theoretical studies just described present a strong argument
for questioning the particularistic perspective. However, that perspective
can be even more effectively challenged if it is scrutinized within the context of its own assumptions. The basis of the particularistic approach is
the supposedly idiosyncratic nature of mortuary rituals; i.e., their inexplicable nature when divorced from their cultural context. With the excep-
tion of Kroeber, the particularists see mortuary ritual as understandable
only in terms of the values and ideology of a particular society (Huntington
and Metcalf 1979). Thus, their argument can be undermined if it can be shown
that, within an individual sociocultural context, variations and changes in
mortuary practices are in part conditioned by recurrent, universal factors,
or that the form of a ritual is determined as much by general as by specific
cultural influences. The remainder of this paper will concentrate on demon-
strating just that.

It is probably a truism (though a frequently overlooked one) that
sociocultural systems of greater complexity tend to allocate not only greater
amounts of energy, but also greater proportions of their energy budgets, to
information processing. Miller (1965:396) has suggested that this allocation
confers a positive survival advantage, and it is not difficult to understand
why this is so. As systems increase in complexity, the need for a coordina-
tion of parts develops disproportionately, leading to such things as increased
numbers of communication channels, increased information volume, specialists
who function in the processing of information, more and more esoteric codes,
and the like. Thus, the costs of information processing are driven up, and
more resources must be committed to it.

Mortuary rituals basically comprise part of a communication system
in which symbols are employed to convey information about the status of the
deceased, the nature of the group participating in the ceremony, and other
factors. As societies increase in complexity and become increasingly hier-
archical, it becomes more and more important to clearly signify functional
roles, levels of authority, channels of command, lines of descent, and
crucial personnel. This is accomplished, as Binford (1962) has pointed out,
by employing an increasing number of physical symbols. It is also accomplished, as Saxe (1970) has emphasized, through mortuary ritual. Within a
given sociocultural context, variations in social complexity generally can
be expected to be reflected not just in different kinds of mortuary symbolism,
but in the overall amount of energy allocated to mortuary ritual. This allocation will not be constrained by cultural context, and is not understandable
in a particularistic framework.

        To illustrate this point, four archaeological examples will be
discussed below. Ethnological examples would make useful comparisons, but
I know of none suitable for our present purposes. Additionally, archaeological data are usually required for controlled studies of long-term social
change.

<u>West-Central Illinois</u>

        The nature of the sociocultural changes that occurred between the
Middle and Late Woodland periods in the Midwest have concerned archaeologists
for many years. The Middle Woodland (or Hopewell) period is widely known for
the degree of cultural complexity that appears to be reflected in the archaeological record. Perhaps the best-known Middle Woodland features involve
systems of elaborate mortuary ritual. Other characteristics include highly
ornate artifact styles, a long-distance trade in exotic raw materials and
perhaps finished artifacts, and (in many locations) the construction of large
earthworks.

        During the Late Woodland period there was a marked curtailment in
mortuary ceremonialism, and a reduction in the elaborateness and diversity of
artifact forms. There was also a reduction in long-distance trade relationships and an end to the construction of extremely large earthworks. The
Late Woodland period is characterized, therefore, by archaeological remains
indicative of a cultural system that had undergone a shift away from earth-
work construction requiring large-scale labor mobilization, away from elaborate
mortuary ceremonialism, and away from certain forms of interaction between
social groups. The archaeological record seems, then, to reflect major cultural changes, specifically a decrease in sociocultural complexity.

        In order to assess the nature of social change during the Woodland
period, a study was undertaken of Middle and Late Woodland mound-group mortuary populations in the lower Illinois River valley of west-central Illinois
(Tainter 1975, 1977). The Middle Woodland period in this area dates between
150 B.C. and A.D. 400, and the Late Woodland period from A.D. 400 to 900. My
analysis of the data focused primarily on energy expenditures in mortuary
treatment. Patterning in the data allowed a delineation of distinctive levels
of energy expenditures. I argued that such distinctive levels of energy
expenditure reflect distinctive levels of social involvement in the mortuary
act, and suggested that grades or levels of ranking could therefore be isolated.

        The study used a sample of 443 Middle Woodland burials from two
mound groups, and 926 Late Woodland interments from five mound groups. The
Middle Woodland data clustered into six levels of energy expenditure and (by
inference) rank.

        <u>Level 1</u>. This level involved individuals placed in, or processed
through, large tombs that formed the central node in most Middle Woodland
mounds. These tombs regularly display log roofs, as well as log or limestone walls, and are surrounded by ramps of loaded earth. Individuals
placed in these tombs were periodically removed and reburied nearby as
bundles of disarticulated bones, while subsequent burials were placed in
the feature. Individuals placed in these tombs were associated with nearly
all of the exotic, imported materials found with mortuary remains. This

mode of disposal reflects greater amounts of energy expenditure than any
other Middle Woodland form of interment.

Level 2. The second level of energy expenditure was reflected in
graves arranged peripherally about the central tomb. These particular graves
had log coverings, but were substantially smaller and less elaborate than
the central log tombs.

Level 3. The third level of energy expenditure involved burials
with limestone slabs placed in the grave, occasionally as specific arrangements around the body.

Level 4. The examples of this level were identified by the placement of certain types of material rank indicators in the grave. The most
common items thus included were elaborate Hopewell-series pottery vessels.

Level 5. The individuals representing this level were interred in
mounds in simple subfloor graves, with little effort being expended on the
construction of the facility.

Level 6. The smallest energy expenditure was allocated to persons
representing this level. These individuals were merely deposited on accre-
tional surfaces of the mound, and small amounts of earth were then piled over
them. This mode of disposal required less effort than the excavation of
subfloor graves for individuals in Level 5.

Within the five Late Woodland data sets, the Schild Mound Group
nicely illustrates the levels of energy expenditure in later mortuary systems;
the other Late Woodland mound groups are basically similar. While the Middle
Woodland mortuary system included six levels of energy expenditure, only five
are evident in the Schild data.

Level 1. The most energy in the Schild mortuary system was expended
on persons buried in large, log-roofed tombs. Some of these tombs were, in
turn, used as crematories. The log tombs in this mound group were substantially less elaborate than those in Middle Woodland mounds.

Level 2. The second level of energy expenditure involved a degree
of custodial care and skeletal manipulation that resulted in the final interment of a disarticulated bundle of bones.

Level 3. Burials representing this level were characterized by
the placement of limestone slabs over the grave.

Level 4. The fourth level in the Schild system involved individuals who were associated with certain kinds of status/rank markers. These
items, like the Late Woodland technological system in general, were noticeably
less elaborate than the grave associations found in Middle Woodland mounds.
While rank indicators in Middle Woodland mounds might include copper ornaments, Gulf Coast shells, and other exotic materials, ornamental grave
associations in Late Woodland mounds are limited to items such as *Anculosa*
shell beads.

Level 5. This level reflects the largest segment of the population
and involves persons who were merely interred in a flexed position in an
otherwise unpretentious grave.

In both the Middle and Late Woodland data sets, the rank levels that have been identified generally display population profiles in which all age grades, from infant to 50+ years, are represented. The expenditure of identical amounts of energy on the interment of both infants and adults indicates that the level of social involvement in the mortuary act was not linked to age-graded distinctions. This suggests that the rank and status positions involved were not acquired by passing through age grades and were not dependent on the performance of age-linked activities. One might interpret these positions, therefore, as largely inheritable (cf. Saxe 1970: 8). There are exceptions to this pattern, and at least a minority of the ranked levels in Woodland social systems involved achieved positions. In all cases, however, the paramount rank level seems to have been inherited and ascribed.

The assessment of social change in this study involved the development of quantitative measures of social complexity, the amount and degree of organization, and rank differentiation. The rationale behind these measures is complex, and the reader is referred to the original study for further details (Tainter 1975, 1977). The use of these quantitative measures yielded results which were congruent with the non-quantitative conclusions reached by earlier investigators. Between the Middle and early Late Woodland periods, there was a general decrease in social complexity, in organization, and in rank differentiation, followed by an increase in these variables in the succeeding Mississippian period.

A comparison of the physical characteristics of the Middle and Late Woodland mortuary systems clearly shows that the less complex society was allocating less energy to mortuary symbolism. In addition, the contrasts are consistent throughout these mortuary systems. Late Woodland burials generally lack the ornate, costly artifacts of shell, copper, obsidian, and the like that Middle Woodland populations obtained at great cost from throughout the eastern two-thirds of the United States and then placed with their dead. The large, log-roofed tombs that held the deceased paramount members of Middle and Late Woodland societies were much more elaborate in the earlier period, displaying such features as fabric wall-coverings and ramps made of earth that was often carried from the river bottom to the bluff-top burial locations. Even among persons of lower rank, there are contrasts in the energy allocated to mortuary processing. Among Middle Woodland populations, the common mode of disposal for those of lower rank was inhumation in an extended position. By the Late Woodland period, economy in mortuary practice dictated flexed interment for such persons. The total energy savings accomplished by this last change was substantial when one considers that in the Schild Mound Group fully 82.4 percent of the population was buried in this manner (Tainter 1977:342).

Before closing this section one point must be clarified. Too quick a reading of the foregoing might lead some readers to think that the argument is circular: the procedures used to infer the occurrence of social changes are also used to validate the argument that such transformations are accompanied by changes in the energy expended in mortuary ceremonialism. Actually, the two conclusions are conceptually and analytically distinct. The inferences regarding social changes are based upon a comparison between the social characteristics of localized groups at different points in time. They monitor changes in structure and organization. They are not tied in *requisite* manner to the symbols used to identify the rank or status of individuals, since many means could be used to express this. Thus, nothing in the data used here would have intrinsically precluded the achievement of a different conclusion.

Central California

One of the few archaeological studies that has approached the topic of social change through mortuary practices was carried out by David Fredrickson (1974), who demonstrated that a shift from egalitarian to ranked social organization occurred between the Middle and Late Horizon in central California. Fredrickson's data span the interval from about 2500 B.C. to the sixteenth century A.D. Fredrickson, who employed an approach derived ultimately from Stickel (1968), showed that Late Horizon burials were more apt to be associated with material symbols that denote rank. Persons buried with such symbols were also frequently singled out for a more costly treatment in the form of cremation.

Overall, the Late Horizon mortuary system appears to have been more costly than that of the Middle Horizon. The use of cremation for high-ranking persons is a clear indication of this. In addition, there was an explosion in the number and variety of beads and shell ornaments which were costly to make and which were relatively scarce in Middle Horizon burials.

Southern California Coast

Somewhat more extensive sets of mortuary data are available for the southern California coast. Three examples which provide a good temporal sequence in addition to sizeable samples are the Rincon site, ca. 1700 B.C. (Stickel 1968), the Fowler site, ca. A.D. 0 (Tainter 1971), and the late prehistoric Medea Creek cemetery (King 1969). The excavators of all three sites employed the theoretical approach developed by Stickel (1968), although additional lines of inquiry were employed for the Medea Creek analysis.

The early Rincon cemetery displayed patterns indicative of an egalitarian society (Stickel 1968). All grave associations were of a utilitarian nature. These did not appear to indicate distinctive statuses, but were quantitatively distributed among burial types. There was no evidence that any infants received a conferred high ranking. Stone features may have been associated with the mortuary locality.

By around 2500 B.P., ranked social systems had developed on the southern California coast. At the Fowler site (Tainter 1971), evidence was found which suggested the presence of two ranked kinship groups, with the one of higher rank utilizing the western segment of the cemetery. The mortuary assemblage had proliferated to include a large array of material rank and status indicators. The diversity and quantity of beads, ornaments, whistles, and other non-utilitarian items contrasted sharply with that discovered at the Rincon site; some burials contained shell beads numbering in the thousands. No burials at the Rincon site had such associations.

By A.D. 1500, native societies in southern California had reached the point where they were among the most complex ever recorded for hunters and gatherers. The Medea Creek cemetery (King 1969) reflects that complexity. This cemetery also suggested the existence of ranked kinship groups, with

the higher ranking group again in the western segment. This western segment
contained lines of reburials, which seem to have served as markers which
demarcated subareas.

The evolution of mortuary complexity (and cost) from the Rincon
site to the Fowler site is clear. The proliferation of non-utilitarian rank
or status markers, sometimes numbering in the hundreds or thousands, at the
Fowler site is not matched by any Rincon burial. However, the continuation
of this pattern cannot be clearly demonstrated at Medea Creek, since the
primary data from this site have not been published. A variety of bead types
and other artifacts denoting differences in rank were nonetheless recovered
from Medea Creek. Some of these artifacts, such as plank canoe fragments,
represent a considerable labor investment. Medea Creek does display some
increased complexity in the treatment of the corpse, which is discernible in
the alignments of reburials. Reburials were rare at the Rincon site, while
none was found at Fowler.

Hawaii

At the time of European contact, the society of native Hawaii was
stratified and complex, probably as complex as is possible for a society that
is not organized as a state. Archaeological analysis has shown, however,
that there were substantial variations in social complexity at the local level
among communities occupied during the contact era (Tainter and Cordy 1977).

The Kaloko cemetery, located on the western side of the island of
Hawaii, contained an assemblage of burials placed in caves, lava crevices,
and on stone platforms. These interments were analyzed according to the
criterion of energy expenditure; this approach was first developed at Kaloko
(Tainter 1973). Six graded levels of platform interment types, followed by
cave and crevice inhumations, appeared to exist at Kaloko. The latter, displaying the least energy expenditure, represented the lowest ranking members
of the community.

At Anaehoomalu, which is also on the west coast of Hawaii to the
north of Kaloko, there was a noticeably different mortuary assemblage.
Anaehoomalu had none of the large, elaborate, labor-consuming platform constructions that dominated the Kaloko cemetery. Of the 85 burials recorded
at Anaehoomalu, 82 were in caves and the others in crevices. Four levels of
energy expenditure were represented by these burials.

Level 1. This level involved individuals who warranted the construction of a stone wall to demarcate their place of interment.

Level 2. This level involved individuals who were buried in, or
associated with, a canoe or canoe parts.

Level 3. This level was represented by disarticulated bundles of
bones.

Level 4. The last level involved individuals who were buried in
an articulated state with no associated stone walls or canoe parts.

The Kaloko and Anaehoomalu data were quantitatively compared by
means of the techniques which were developed during the study of Woodland

social changes (Tainter 1975, 1977); i.e., the two areas were quantitatively
compared for structural complexity, amount and degree of organization, and
degree of rank differentiation. All measures indicated that Kaloko represented a more complex, more highly organized, and more highly stratified
society than Anaehoomalu. It is not surprising, therefore, that the Kaloko
mortuary assemblage reflects a much greater energy investment than Anaehoomalu,
for the latter contains none of the expensive platform constructions of the
former.

The Kaloko and Anaehoomalu cases are important for a number of
reasons. Both communities were contemporaneous segments of the same sociopolitical entity, whereas in the previous three cases, the comparisons
involved changes through time. These two communities, therefore, almost
certainly reflect compatible, if not nearly identical, symbolic systems.
In addition, conclusions concerning the relative social characteristics of
the two communities are corroborated by an independent set of data. Analysis
of settlement patterns at Kaloko and Anaehoomalu revealed the same phenomenon
reflected in the mortuary remains; the former community was more complex and
more highly stratified than the latter (Tainter and Cordy 1977).

CONCLUDING EVALUATION

The particularistic approach to the study of mortuary practices has
been characterized by a variety of perspectives, ranging from Kroeber's (1927)
assertion that mortuary practices do not form a systemic linkage with other
cultural phenomena, to Huntington and Metcalf's (1979) view that mortuary
behavior can only be understood in a cultural context. The particularists
are united, however, in their view that the pursuit of generalizations,
"elusive cultural universals," is unprofitable if not impossible.

As part of an attempt to empirically refute this view, four case
studies have been examined. In all four cases, it is possible to argue that
the *forms* of mortuary symbolism remained fairly constant. In the Woodland
example, mortuary symbolism involved the form of the interment facility
(mounds, walled tombs), the handling of the corpse (reburial, cremation,
extension/flexion), and (to a limited degree) the nature of grave associations.
In the two California cases the main variations in mortuary treatment involved
the inclusion of various kinds of beads, ornaments, and other rank or status
markers in the graves. Cremation also served as an indicator of social differentiation in Late Horizon central California. The Woodland and California
cases thus show essential constancy in mortuary representation through time.
In Hawaii, the two communities compared were members of the same cultural
and sociopolitical entity, and thus presumably shared a common symbolic
repertoire.

The examples have demonstrated two things. First, the notion that
mortuary symbolism is exclusively culture-specific, and that the form of a
mortuary ritual can only be understood in context, has been contradicted.
Mortuary rituals do not vary idiosyncratically. To a large extent, their
distribution is constrained along dimensions of social complexity and energy
expenditure. The more complex a society, the more energy it must allocate
to information processing, including information concerning status obligations

to the deceased. This requirement will persist regardless of cultural context, and will profoundly influence the symbolism employed in mortuary
ritual.

Second, the contention that the study of mortuary practices cannot
produce generalizations has been shown to be without merit (Saxe 1970;
Binford 1971; Tainter 1973, 1975; Goldstein 1976). Contrary to the opinions
of Kroeber (1927), Ucko (1969), and Huntington and Metcalf (1979), this realm
of human behavior is no less amenable to objective study than any other. It
is time for both ethnologists and archaeologists to include mortuary practices in the generalizing world of science.

5 The Role of Facilities in Technological Systems

Michael A. Glassow

INTRODUCTION

   A useful objective of a systems approach in cultural anthropology
is the development of concepts for more effectively characterizing the nature
of cultural systems. Indeed, the power of explanations of cultural variability and change is largely dependent on the clarity of the concepts used
to express them. That is, a conceptual structure used to characterize
aspects of cultural systems must have specific reference to discrete, iden-
tifiable system parts and the ways in which these parts are linked together
in matter-energy exchanges. With this lesson in mind, we may consider concepts that refer to the manner in which items of technology are used by humans
to manipulate matter and energy. While archaeologists have traditionally been
concerned with technology, a theory to guide the study of technological systems is still very poorly developed. One of the most useful discussions of
technological systems is found in Philip Wagner's *The Human Use of the Earth*
(1960). In an effort to derive meaningful units found in a means of production, Wagner proposes two elementary classes of technological items: *tools*
and *facilities*. His definitions are given here.

   Tools, according to Wagner, are "movable material objects especially
designed for application of energy in precise and controlled ways for particular mechanical tasks" (1960:94). In other words, "the tool operates as an
extension of man's own body, and transmits the motions of the body" (1960:
94). *Facilities*, on the other hand, "are stationary artificial objects whose
function is to contain or restrain motion" (1960:94).

   My concern here will be exclusively with Wagner's concept of
facility. I shall first define two different kinds of facilities, and then
present some thoughts on how the use of facilities relates to economy.
Finally, I shall discuss the relevance of the concept of facility in propos-
ing explanations of technological change.

TYPES OF FACILITIES

   Wagner divides facilities into several formal categories. *Containers*
include vessels of various types, buildings, tubes, and simple watercraft.
Bases are two-dimensional surfaces such as floors, roads, or food-drying
platforms; *barriers* include such items as dams, fences, and walls. A final
category is *lines*, which consists of cords, cables, and the like (1960:97-

99). Most technologies that are based primarily on human energy as a source
of power are composed of simple tools and facilities. Wagner recognizes, however, that even the simplest technologies do contain more complex artifacts
which are, in essence, combinations of simple tools and facilities. Most of
these would be primitive forms of *machines* in Wagner's classification (1960:
98). The bow-and-arrow and mortar-and-pestle are examples. In fact, Wagner
has developed a typology of technological artifacts that is meant to embrace
the full range of complexity found in modern industrial technologies.

Wagner's classification of facilities into containers, bases,
barriers, and lines emphasizes the formal properties of facilities. Another
approach to classification--one that would solve some of the inconsistencies
in his classification--might emphasize the role of facilities in allowing
human populations to manipulate their matter-energy resources. In this
regard, facilities may be seen as being used in two basic processes: *storage*
and *conversion* of matter or energy. In the first process, the objective is
to keep a resource in a relatively constant state; in the second, the objective
is to convert the resource from one state into another. The latter form of
facility is what Wagner calls an *energy converter*, which he places on the same
taxonomic level as tools and facilities. Wagner fails to recognize, however,
that some energy converters, such as roasting pits or cooking pots, are
actually simple facilities in terms of his own definition. Both types of
facilities accomplish their designated processes through the creation of what
Wagner has called *artificial environments* (1960:8, 21, 22, 118), which are
human-created environments conducive to the preservation or conversion of
matter or energy. By partitioning segments of the natural environment with
facilities, human populations have been able to exclude certain natural forms
of matter or energy while including certain others, thus stopping or slowing
down certain natural processes of matter or energy disintegration or enhancing
others.

Hunter-Anderson (1977:295) has recently attempted a different type
of classification of what she refers to as "passive" facilities, utilizing
the criterion of whether the principal role of the facility is preservation
or protection. *Houses*, she claims, serve primarily to protect their contents
from at least some aspects of the external environment. *Containers*, in contrast, prevent the dispersion of contents. However, it would be difficult
to find facilities that serve one of these functions exclusively; many, in
fact, appear to serve both functions equally--and sometimes alternatively,
as when a house serves to contain heat energy in winter but withhold it in
summer. It would be more appropriate to say that Hunter-Anderson has identified two ways in which storage facilities control the flow of matter or
energy.

Another approach to classification used by Hunter-Anderson is more
operational. She notes that some passive facilities are meant to contain
undifferentiated contents, such as water or grain, whereas others are meant
to contain differentiated and often variably-shaped contents, such as tools
in a tool chest or clothes in a chest of drawers. She notes that the former,
which she calls *bins* or *binhouses*, tend to have round shapes whereas the
latter, called *wareboxes* or *warehouses*, tend to have rectangular shapes
(Hunter-Anderson (1977:296-302).

Facilities may be said to maximize the utility of the various forms
of matter or energy upon which a human population depends. This is

accomplished by extending the time and/or space in which matter-energy
resources are available to a human population; i.e., *time utility*, *space
utility*, and *form utility* are obtained from these resources (Wagner 1960:
120). The preserved storage of a food product, for example, will make it
available considerably beyond the period of time the product is naturally
fresh, and the facilities used in transportation make a food product available over a much larger region than that from which the product originates.
Conversion facilities give form utility to resources; that is, they convert
resources from relatively unusable to usable forms (Wagner 1960:120). For
instance, a cooking pot may be used to convert cereals from indigestible to
digestible forms.

THE EFFICIENCY OF FACILITIES

With regard to the flow of matter and energy, both storage and
conversion facilities are cultural system components with specific forms,
quantities, and rates or periods of *inputs* and *outputs*. With respect to
storage facilities, we may speak of the input of matter or energy for storage
and, of course, the output of this energy after the storage period. Conversion facilities normally involve at least two different kinds of inputs:
that which is to be converted and that which effects or catalyzes the conversion. Both storage and conversion processes result in some matter or
energy loss at relatively high states of entropy, so we may conceive of a
facility as having a certain *efficiency*, which in simplest terms is the proportion or the amount (or rates) of output to input. This is expressed
diagrammatically in Figure 5.1. Conversion facilities may be thought of as

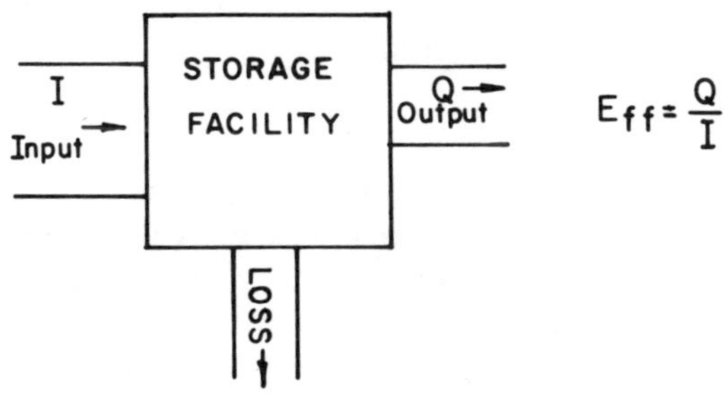

$$E_{ff} = \frac{Q}{I}$$

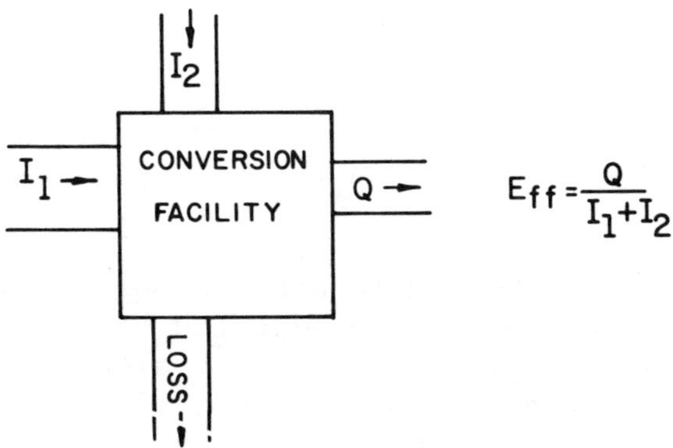

$$E_{ff} = \frac{Q}{I_1 + I_2}$$

Fig. 5.1. Efficiencies of facilities

having an additional type of efficiency: *conversion efficiency*. Here we are
interested in the amount of energy it takes to produce the conversion. The
smaller the input effecting the conversion (I_2) is in proportion to the energy
output ($Q + \text{loss} = I_1 + I_2$), the more efficient the facility is in the conversion process. The formula for conversion efficiency is:

$$E_c = \frac{I_1}{I_1 + I_2} \, .$$

To measure the efficiencies of facilities, or simply the quantity
or rates of flow of the matter or energy processed by facilities, we must
have some sort of common unit of measurement. Actually, there are a variety
of ways for directly or indirectly measuring matter and energy. For
instance, foods and fuels and human effort are often measured in terms of
caloric value, whereas metals or ceramic materials are typically measured as
mass, weight, or volume of matter. We may find that the volume or capacity
of the facility may act as an indirect measure of the matter or energy processed by it, particularly if we are comparing two or more different facilities and can assume that the rates of matter or energy processed by each are
the same.

There is more to the efficiency of facilities than that connected
with their storage or conversion of matter-energy. The efficiency of *space
utilization* within the facility may be considered since some facility shapes
will allow much closer packing of contents than others, especially if the
contents are differentiated. The efficiency of *access* may also be important.
Among the more obvious dimensions of access are rates or frequencies of inputs
and outputs and the sizes and shapes of the units put into or taken out of
the facility. As Hunter-Anderson (1977:297-300) demonstrated, the shape of
a facility will be designed so as to minimize costs of access and maximize
space utilization.

Efficiency may also be seen in the energy costs involved in creating, maintaining or repairing, and replacing facilities. Seldom, if ever,
are forms found already existing in nature specifically suited to serve as
facilities without further modification. The caves or rock shelters that
were occupied prehistorically in various regions of the world probably come
closest to this ideal situation. Nevertheless, it seems reasonable to assert
that humans modify nature as little as possible in creating facilities. They
either consume their metabolic energy or their fuels in *transporting* raw
materials to the site of the facility; they then consume more energy in using
these raw materials in *manufacturing* the facility. Throughout the course of
existence of that facility, they must consume more energy--and very likely
gather more raw materials--in *maintaining* or repairing the facility, and (if
the facility can no longer be repaired) more energy and materials are needed
to *replace* it. The processes involved in the life of a facility may be conceived of as a series of steps, as shown in Figure 5.2.

The potential costs involved in creating or replacing a facility
will, of course, be evaluated against the expected benefits. Many of these
benefits, however, are difficult to identify, let alone quantify, because
they are realized only after considerable periods of time. Benefits may be
seen in a provision for, or an increase in, the three utilities discussed
earlier. In addition, benefits of replacement may be related to decreases
in the costs of construction, as well as increases in life-span.

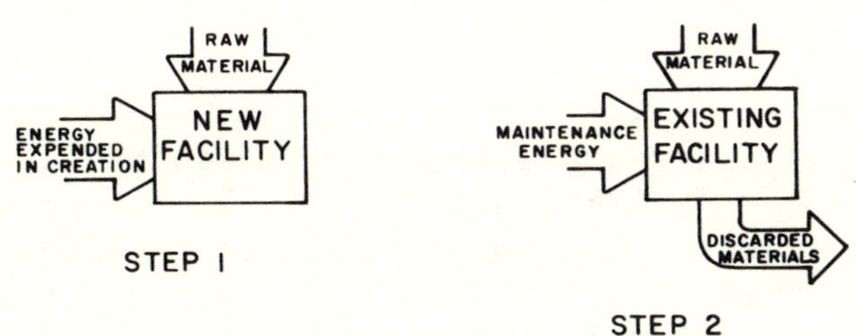

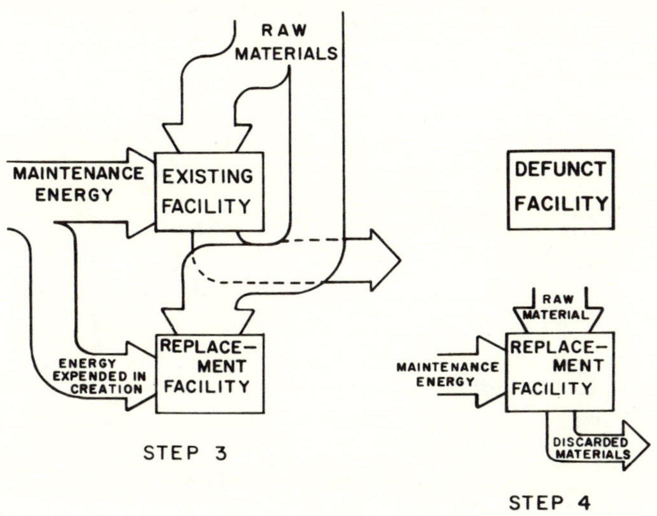

Fig. 5.2. The life of a facility

EXPLANATION OF VARIATION IN THE FORM AND USE OF FACILITIES

Although the concept of the facility is not widely recognized, a growing number of archaeologists have nevertheless identified specific types of facilities and have discussed their importance in understanding the nature and development of technological systems. Some archaeologists have noted the role of certain kinds of facilities at particular junctions in cultural evolution. Binford (1968b:333), for instance, has pointed out that Mesolithic adaptations were oriented around the use of food storage facilities. Flannery (1969:78) has also discussed the importance of facilities among Mesolithic seed-gatherers of the Near East, and referred to the use of granaries as a "pre-adaptation" preceding the development of agriculture. Indeed, when it is realized that agriculture depends to a large extent on the use of facilities, its origin in the Near East, and perhaps also in Mesoamerica, was not the profound "revolution" that either implicitly or explicitly it has been thought to have been. All of the principal facilities of an agricultural technology--such as granaries, grinding implements, and pottery and basketry containers--were already important in Mesolithic or Archaic adaptations prior to the inception of agriculture. The only new facility associated specifically with plant domestication was the agricultural field--a conversion facility (or container in Wagner's terminology) from which naturally-growing plants were excluded and within which seeds of domesticated plants were planted and harvested. With the introduction of soil and water-control devices (e.g., terraces and irrigation ditches) and fertilizers, the characterization of the agricultural field as a type of conversion facility is even more apparent.

Binford has also discussed the role played by ceramics when human populations became more sedentary and eventually began to depend upon an agricultural subsistence base (Binford and Chasko 1976:138-139). He notes that ceramics arose when stored foods coming in "small packages" had to be prepared--in particular, by means of boiling. He suggests that this form of food preparation may have allowed children to be weaned earlier and mothers to bear offspring more frequently. Thus, he concludes that "other things being equal, we might expect increased rates of population growth in response to increased realized fertility to follow the adoption of ceramics . . ." (Binford and Chasko 1976:139).

Schalk (1977:232) has discussed the place of facilities in the development of a dependence on anadromous fish. He argues, in fact, that the use of facilities such as weirs, drying racks, boxes, and storage pits is crucial if resource specialization involving anadromous fish is to occur. Schalk's observations demonstrate that the development of agriculture and the dependence on stored seeds and nuts that preceded it are not the only contexts in which facilities play a significant role in cultural evolution.

The study of facilities may also be concerned with geographical variations in the use of facilities, as demonstrated in Binford's (1978) consideration of the conditions under which meat storage would be expected. He notes that the climatic factor of temperature plays a significant role in determining whether meat will be stored. Drying meat for storage would be expected in those more northerly latitudes of the world where storage over lean seasons becomes profitable, and it would be expected to increase as growing seasons become shorter. Further, the processing time required to prepare meat for storage would decrease with distance from the equator (Binford 1978:93). The Nunamiut Eskimo, living in an extreme northern latitude, may depend more on stored food resources than any other documented group of hunter-gatherers, according to Binford (1978:458). As a corollary to Binford's generalizations, we can expect that meat-storage facilities will generally become more important with distance from the equator.

While it is often obvious that certain features of the archaeological record are indeed facilities--if only because they are containers--their use in processing a specific resource is not always so evident. Puleston's (1971) study of Classic Maya "chultuns," underground chambers found at sites throughout the Guatemalan Peten, is a case in point. While one form of these chultuns that is prevalent in the Yucatan is almost surely a cistern for the storage of fresh water, those of the Peten, which differ somewhat in shape and size, appear to have been used for the storage of some sort of food product. His experiments indicate that maize, beans, and squash could not be stored successfully in these chultuns but that ramon nuts, from trees native to the region, preserved very well for many months.

Other workers have been interested in facilities as measures of the rates of flow of particular resources through a cultural system. For example, Schiffer (1972) attempted to determine the relative importance of agriculture to the subsistence of early farmers of the American Southwest (circa A.D. 700) by measuring the capacity of storage facilities associated with individual households. In essence, he used storage capacity as a reflection of the rate of consumption of agricultural products. The capacity of facilities has also been used to measure the size of social groups, as

exemplified by Naroll's (1962) house-floor index and Turner and Lofgren's
(1966) use of size and abundance of pottery containers as an indicator of
household-group size.

The above references to facilities have taken two quite different
forms. Some scholars have acknowledged that certain kinds of facilities were
used, invented, or changed form during the course of prehistory, and in some
instances changes in facilities have been included in general explanations of
cultural change. Other scholars have depended upon relationships between
formal properties of facilities--particularly size--and other cultural variables in order to use these properties as measures. None of these studies
has devoted much attention to the explanation of specific changes or variations in facilities.

Hunter-Anderson's insightful discussion of storage facilities is
unique in that it does address the problem of explaining differences in forms
of facilities. Her explanation of differences in shape between bins and binhouses on the one hand and wareboxes and warehouses on the other with reference
to the nature of the contents they are meant to store has already been men-
tioned. She adapts this contrast to explain differences in house form. Round
houses, she argues, are associated with few and relatively undifferentiated
household activities, with low volumes of materials and facilities stored
within. However, when household activities become greater in number and more
differentiated in time and space, and when higher volumes of materials and
facilities must be stored, there will be a selection for elongated or recti-
linear floor plans, especially when greater numbers of separate activities
must be performed simultaneously.

Clearly, there is much to be gained from a closer attention to
variations in the form and use of facilities. In the first place, we should
be able to arrive at a much more profound understanding of the nature of cultural change when explanation is focused on variation in specific components
of a cultural system--in this case, a particular kind of facility. Second,
we will be able to develop much more powerful measures of social, subsistence,
and other cultural variables which are linked to the use of facilities when
the determinants of variation in a given class of facilities have been clearly
identified. My objective in the rest of this paper will be to demonstrate
some of the ways in which the concept of facility may be used in the development of explanations of technological change. To this end, it will first be
necessary to specify the components of a model from which hypotheses accounting for change in facilities are to be generated. This model will deal
specifically with those facilities which are directly articulated with the
storage and processing of resources.

The first component of such a model might be the citation of the
ultimate causes of change, which must be sought outside the cultural system.
These ultimate causes may be grouped into two general categories: environmental and demographic. Each concerns the manner in which a population
relates to its environment, and it may be argued that both environmental and
demographic change will inevitably affect either the type or the manner of
exploitation of resources. Nonetheless, this component of an explanatory
model is optional. It could be included if the interest of the investigator
is primarily ecological and excluded if the interest is more strictly ethnological.

The second component is the specification of changes in resource exploitation that have occurred, or that are required, given environmental or demographic change. Changes in resource exploitation may take one or a combination of three different forms. First, there may be a change in the rate at which a resource is processed, as when a resource becomes more scarce in the environment or when it becomes economically more advantageous to increase dependence on a resource. Second, there may be a change in the type of resource exploited; i.e., one resource may be dropped from the repertoire and replaced by another that fulfills the same matter or energy requirements. Third, there may be a change in the efficiency requirements associated with the exploitation of a given resource. Strong economic incentives may arise, for instance, favoring an increase in the product yield from a given unit value of raw resource. Conversely, such economic incentives as already exist may be reduced, implying that a given level of efficiency in production is no longer necessary or economical.

The specification of the effects on existing facilities of the changes in patterns of resource exploitation comprises the third component. These effects may be grouped into two categories: the processing capability (including degree of access) of the facilities may be diminished, or the construction, maintenance, or use-costs may increase. The first involves reductions in the facility's ability to produce or provide output in the required form, quantity, or rate, whereas the second involves the amount of energy required to produce the output.

The final component deals with the modifications in the facilities that are necessary in order to process a new form or quantity of a resource. There are three logical alternatives, the first of which involves no change in the facility at all, but rather a change in the way the facility is used. For instance, the intensity of use of a facility may increase or decrease, which would only affect the replacement rate of the facility. The other two possibilities are: (1) modification of the existing facility so that its capacity is larger or smaller; or (2) revision of the form of the facility so that it more effectively meets new requirements. Both changes may involve either altering existing facilities or replacing them with completely new facilities.

To illustrate how careful attention to these components of a model dealing with changes in facilities can result in the generation of some interesting hypotheses concerning technological change, I turn now to a specific example. I have chosen for this example the shift from underground storage pits to aboveground granaries--a change in the form of binhouses in Hunter-Anderson's (1977) terminology--that took place around A.D. 700 in many parts of the northern portion of the Southwestern United States. This shift was basically one involving form. The earlier of the two forms consisted of a subterranean pit of a meter or more in depth with a constricted opening, usually no more than a half-meter in diameter, that gave the pit the overall shape of a squat bottle. Where the substrate was very rocky or sandy, the pits were normally semisubterranean, with the ground portion having a slab-rock lining. The later form was an aboveground masonry or adobe-walled room, usually rectangular in shape, with an entrance in either a wall or the flat roof.

     An appropriate way to begin developing hypotheses to explain the
shift in form is to ask what was processed through each of these facility
types. As it turns out, there does not appear to have been any significant
difference; both were used to store maize, and although the variety of maize
may have changed through the course of time, there does not appear to be any
good reason to believe that storage requirements were different for the
varieties. This leaves two other possibilities: a change in either the
amount of storage capacity per capita population or a change in the storage
efficiency requirements. Since there is no obvious reason to quickly reject
either possibility, we have the basis for developing two alternative hypotheses.

     There is considerable evidence that storage capacity requirements
did indeed increase through the course of the shift from underground to aboveground storage units (Glassow 1972, 1980), so there is an empirical justifi-
cation for hypothesizing that an increase in storage capacity requirements
was a determinant of this shift. On the other hand, the principal determinant
could have been stresses favoring greater storage efficiency. Underground
pits may be more prone to penetration by rodents or moisture than aboveground
granaries, and as stored maize became increasingly more important in subsistence, the occasional loss of a bushel or two was less tolerated. At
present, the data to allow one of these two alternatives to be rejected in
favor of the other are not available.

     Our fleshing-out of alternative hypotheses must not stop here, how-
ever. We must specify how increases in capacity or efficiency requirements
affected the utility of underground storage pits, and in what way aboveground
granaries were improvements.

     If we played the role of devil's advocate, we might wonder why
increases in storage capacity per capita population did not simply result in
an enlargement of the size of the underground pits in order to accommodate
more stored maize. This may not have been a feasible alternative, especially
if deep adobe soil beneath a well-drained landform was not locally available.
For that matter, an adobe soil may not have the structural properties to allow
large bottle-shaped pits with small openings to hold their shape for very
long.

     So one branch of the hypothesis that a change in storage facility
form was due to increased storage needs would argue that an increase in stored
maize needed per capita population required an abandonment of underground
storage chambers simply because large underground chambers could not be built,
at least in certain habitats. An alternative branch would argue that the
amount of labor required to construct an underground chamber increased geo-
metrically as capacity increased, and that overall costs associated with the
(successful) storage of a bushel of maize was eventually reduced by adoption
of aboveground granaries.

     The hypothesis citing a stress for greater storage efficiency as
a cause has three possible branches. The first of these is that the switch
to aboveground storage came about when increasing storage-capacity requirements resulted in aboveground storage becoming more efficient as a consequence
of a reduction in losses from spoilage or predation. The second branch is
that access efficiency would have decreased with an increased size in underground storage chambers. Although access to aboveground granaries appears

to have resembled access to storage pits (in that both are entered from the
top by means of a narrow hole), the structural properties of the adobe soil
may have required longer entrance necks to larger storage chambers. Thus,
the difficulties involved in getting in and out of a storage pit would have
increased significantly. The third branch is more complicated. It argues
that construction and perhaps maintenance costs would rise very sharply if
underground storage pits were improved for greater efficiency. Aboveground
granaries would therefore allow greater storage efficiency at lower construction or maintenance costs. This branch, incidentally, does not depend upon
an assumption of increasing storage capacity during the course of the shift
in form.

As a result of a consideration of the two alternative stresses on
underground storage facilities and the different ways in which these stresses
affected their use, we have generated five alternative hypotheses. This process of hypothesis formation is illustrated in Figure 5.3. Available evidence
does not allow us to reject any one of these hypotheses at this time. Indeed,
it is possible that two or more of the alternatives may be important, or that
the relative importance of each varies with regional environmental conditions.

While I cannot offer a clear solution to the problem of why the
form of storage facilities shifted from underground pits to aboveground rooms,
I have at least been able to demonstrate the value inherent in being concerned with the formal properties of a facility type, the manner in which
resources were processed by it, and the economics associated with its use.
When the focus is as specific as in this example, explanation becomes the
complex endeavor of attempting to eliminate a series of competing hypotheses
from further consideration. While I have not gone into the possible test
implications of each of the hypotheses, it is apparent that the data necessary to test them will be of kinds not often collected by archaeologists.
Obviously, an experimental simulation of the type undertaken by Puleston
(1971) in his attempt to identify the use of Classic Maya chultuns would be
a relevant avenue of research. In the course of such research, data must be
collected on the economics of facilities use, including manufacturing and
maintenance costs and storage efficiency (the latter was one of Puleston's
major concerns).

SUMMARY AND CONCLUSIONS

In this paper I have adapted Philip Wagner's concept of facility
and refined its definition so that it might better apply to a discrete realm
of technology. In doing so, I have defined two types of facilities:
storage and conversion facilities. I have also attempted to outline the
beginnings of a theoretical structure that could be used in studying the
nature and use of facilities. Finally, I have attempted to demonstrate
through example the utility of studying the manner in which facilities are
used to process resources, in order to explain their changes in form through
time.

The study of the way in which facilities are used may be construed
to be an example of the "systems approach" in archaeology, since its concern
is with the manner in which facilities--components of a larger technological
system--are used to process certain forms of matter or energy (or what we

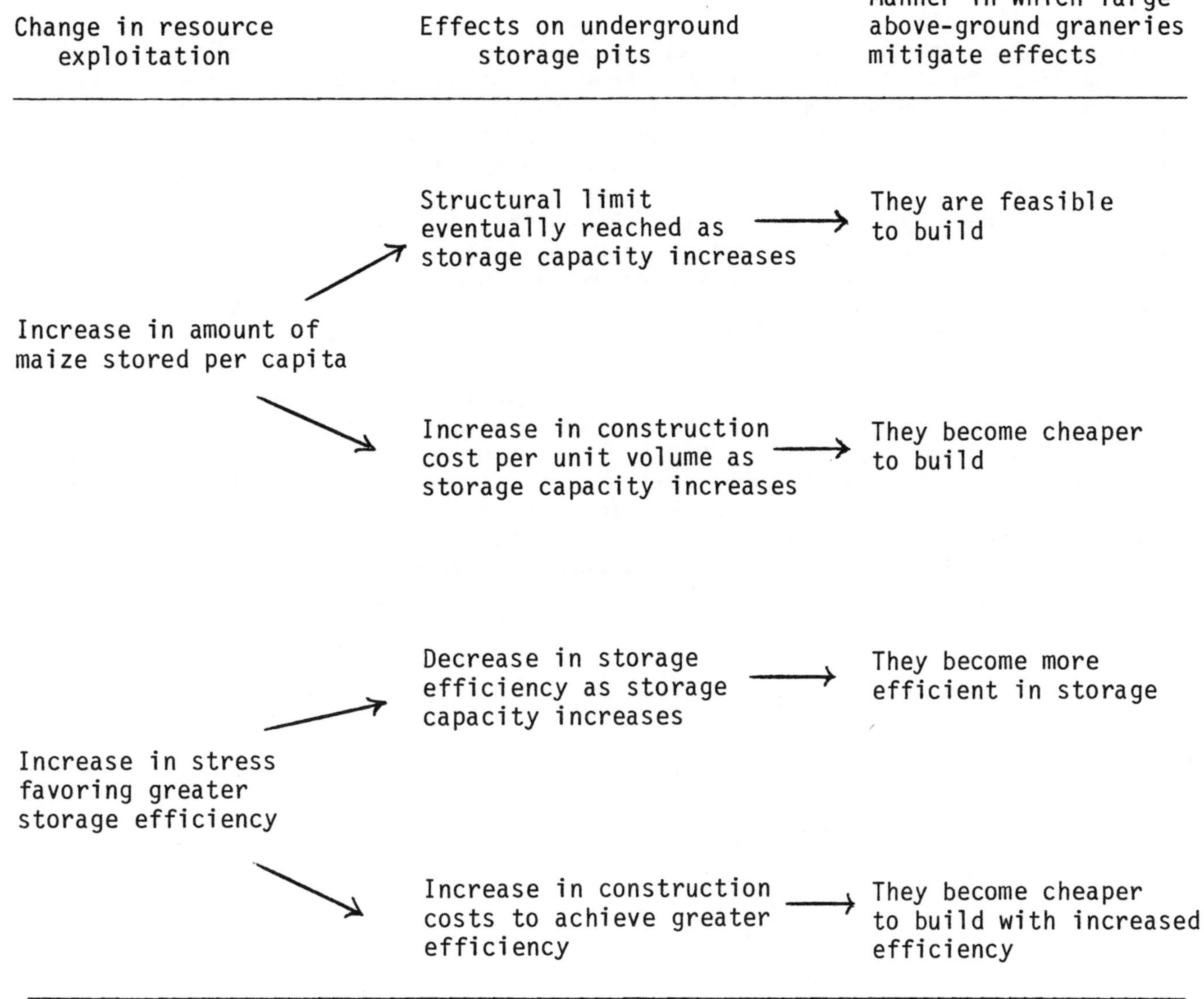

Fig. 5.3. A branching diagram illustrating the generation of hypotheses accounting for change in Southwestern storage facilities

normally call "resources" in archaeology). Although systems analysis--or
more specifically, the use of systems concepts--has gained an unsavory
reputation among some circles of anthropologists (Doran 1970; Spaulding
1973:345-347), it is worth pointing out that any responsible explanation
(which implies the use of general laws) of cultural variability in anthropology really involves systems analysis (LeBlanc 1973:204-214; Spaulding
1973:344). The principal difference between the sound explanations of those
who call themselves systems analysts and of those who cast a suspicious eye
at systems analysts is in the language used in the explanations. The dis-
paragers of systems analysis argue that systems language is largely unneces-
sary jargon (Service 1969:74) or is misused by archaeologists (Doran 1970:
294), and indeed they are partly right. On the other hand, most archaeologists
who have adopted the language of the systems approach believe that they are
really adopting concepts that are more useful in characterizing parts and processes of cultural systems than those that already exist in archaeology and
cultural anthropology. Insofar as this is actually the case, I subscribe to
this borrowing, and my presentation of the concept of the facility is meant
to be just this. Perhaps the use of systems concepts has been initially very
crude, but this appears to me to be simply a result of the rather impoverished
state in which the scientific study of human behavior still exists. If it is
done carefully, the borrowing of systems concepts and, more broadly, the use
of the systems approach, very often allows us to express explanatory problems
that could neither be conceived of nor expressed using the traditional conceptual structure of archaeology or cultural anthropology.

REFERENCES

Adams, R. McC.
  1966   *The Evolution of Urban Society.* Chicago: Aldine.

Barth, F. (ed.)
  1969   *Ethnic Groups and Boundaries.* Boston: Little, Brown.

Bendann, E.
  1930   *Death Customs: An Analytical Study of Burial Rites.* London:
          Keagan Paul, Trench, Trubner.

Berry, C. F.
  1967   *Geography of Market Centers and Retail Distribution.* Englewood
          Cliffs: Prentice-Hall.

Binford, L. R.
  1962   Archaeology as Anthropology. *American Antiquity* 28(2):217-225.

  1963   Red Ochre Caches from the Michigan Area: A Possible Case of Cul-
          tural Drift. *Southwestern Journal of Anthropology* 19:89-108.

  1964   A Consideration of Archaeological Research Design. *American
          Antiquity* 29:425-441.

  1965   Archaeological Systematics and the Study of Cultural Process.
          American Antiquity 30:203-210.

  1968a   Archaeological Perspectives. In *New Perspectives in Archaeology,*
          L. R. Binford and S. R. Binford (eds.), pp. 18-23. Chicago:
          Aldine.

  1968b   Post-Pleistocene Adaptations. In *New Perspectives in Archaeology,*
          L. R. Binford and S. R. Binford (eds.), pp. 313-341. Chicago:
          Aldine.

  1971   Mortuary Practices: Their Study and Their Potential. In *Approaches
          to the Social Dimensions of Mortuary Practices,* J. A. Brown (ed.),
          pp. 6-29. *Society for American Archaeology Memoirs* 25.

  1972   *An Archaeological Perspective.* New York: Academic Press.

  1978   *Nunamiut Ethnoarchaeology.* New York: Academic Press.

Binford, L. R. (ed.)
  1977a   General Introduction. In *For Theory Building in Archaeology:
          Essays on Faunal Remains, Aquatic Resources, and Systemic Model-
          ing,* L. R. Binford (ed.), pp. 1-10. New York: Academic Press.

Binford, L. R. (ed.) [cont'd.]
   1977b   *For Theory Building in Archaeology: Essays on Faunal Remains,
           Aquatic Resources, and Systemic Modeling.* New York: Academic
           Press.

Binford, L. R., and S. R. Binford (eds.)
   1968   *New Perspectives in Archaeology.* Chicago: Aldine.

Binford, L. R., and W. J. Chasko, Jr.
   1976   Nunamiut Demographic History: A Provocative Case. In *Demographic
          Anthropology*, E. B. W. Zubrow (ed.), pp. 63-143. Albuquerque:
          University of New Mexico Press.

Bökönyi, S.
   1974   *History of Domestic Mammals in Central and Eastern Europe.*
          Budapest: Akadémiai Kiadó.

Bray, W.
   1972   The Biological Basis of Culture. In *The Explanation of Culture
          Change: Models in Prehistory*, C. Renfrew (ed.). London: Duck-
          worth.

Brothwell, D., and E. S. Higgs
   1970   *Science in Archaeology* (2nd ed.). London: Thames and Hudson.

Brown, J. (ed.)
   1971   Approaches to the Social Dimensions of Mortuary Practices.
          Society for American Archaeology Memoirs 25.

Campbell, B. G.
   1976   *Humankind Emerging.* Boston: Brown.

Chadwick, J.
   1961   *The Decipherment of Linear B.* Baltimore: Penguin.

Childe, V. G.
   1957   *The Dawn of European Civilization.* London: Routledge and Kegan
          Paul.

Christenson, C., and D. Read
   1977   Numerical Taxonomy, R.-Mode Factor Analysis, and Archaeological
          Classification. *American Antiquity* 42:163-179.

Clark, J. G. D.
   1972   Star Carr: A Case Study in Bioarchaeology. *Addison-Wesley
          Module in Anthropology* 10.

Clarke, D. L.
   1968   *Analytical Archaeology.* London: Metheun, Ltd.

   1972a   Models and Paradigms in Contemporary Archaeology. In *Models in
           Archaeology*, D. L. Clarke (ed.), pp. 1-6. London: Metheun.

Clarke, D. L. (ed).
   1972b   *Models in Archaeology.* London: Metheun.

Clarke, D. L., and R. Chapman
   1978   *Analytical Archaeology* (2nd ed.). New York: Columbia University
           Press.

Cleland, C. E.
   1972   From Sacred to Profane: Style Drift in the Decoration of Jesuit
           Finger Rings. *American Antiquity* 37:202-210.

Cleland, C. E. (ed.)
   1976   *Cultural Change and Continuity.* New York: Academic Press.

Deetz, J.
   1967   *Invitation to Archaeology.* New York: Natural History Press.

   1972   Settlement Patterns in New England: 17th-19th Centuries. Paper
           presented at the Archaeological Institute of America Meetings,
           New York, New York.

DiPeso, C.
   1974   *Casas Grandes: A Fallen Trading Center of the Gran Chichimeca.*
           Flagstaff: Northland Press.

Doran, J.
   1970   Systems Theory, Computer Simulations and Archaeology. *World
           Archaeology* 1(3):289-298.

Dunn, F.
   1970   Cultural Evolution in the Late Pleistocene and Holocene of South-
           east Asia. *American Anthropologist* 72:1041-1054.

Dunnell, R. C.
   1978   Style and Function: A Fundamental Dichotomy. *American Antiquity*
           43(2):192-202.

Earle, T. K., and J. Ericson (eds.)
   1977   *Exchange Systems in Prehistory.* New York: Academic Press.

Fagan, M.
   1975   *In the Beginning: An Introduction to Archaeology* (2nd ed.).
           Boston: Little, Brown.

Flannery, K. V.
   1968   Archaeological Systems Theory and Early Mesoamerica. In *Anthro-
           pological Archaeology in the Americas*, B. J. Meggars (ed.), pp.
           67-87. Washington, D.C.: The Anthropological Society of Wash-
           ington.

   1969   Origins and Ecological Effects of Early Domestication in Iran and
           the Middle East. In *The Domestication of Plants and Animals*,
           P. J. Ucko and G. W. Dimbleby (eds.), pp. 73-100. Chicago:
           Aldine.

   1972   The Cultural Evolution of Civilizations. *Annual Review of Ecology
           and Systematics* 3:399-426.

Foucault, M.
1972   *The Archaeology of Knowledge and the Discourse of Language.* New
        York: Harper Colophon Books.

Fredrickson, D. A.
1974   Social Change in Prehistory: A Central California Example. In
        'Antap: California Indian Political and Economic Organization,
        L. Bean and T. King (eds.), pp. 57-73. Ramona: Ballena Press.

Fried, M. H.
1967   *The Evolution of Political Society.* New York: Random House.

Friedrich, M. H.
1970   Design Structure and Social Interaction: Archaeological Impli-
        cations of an Ethnographic Analysis. *American Antiquity* 35:332-
        343.

Fritz, J. M.
1978   Paleopsychology Today: Ideational Systems and Human Adaptation
        in Prehistory. In *Social Archeology: Beyond Subsistence and
        Dating,* C. L. Redman et al. (eds.). New York: Academic Press.

Fritz, J., and F. Plog
1970   The Nature of Archaeological Explanation. *American Antiquity* 35:
        405-412.

Glassow, M. A.
1972   Changes in the Adaptation of Southwestern Basketmakers: A Systems
        Perspective. In *Contemporary Archaeology,* M. P. Leone (ed.),
        pp. 289-302. Carbondale: Southern Illinois University Press.

1980   *Prehistoric Agricultural Development in the Northern Southwest.*
        Socorro: Ballena Press.

Goldstein, L.
1976   *Spatial Structure and Social Organization: Regional Manifesta-
        tions of Mississippian Society.* Ph.D. dissertation, Department
        of Anthropology, Northwestern University.

Goodenough, W. H.
1965   Rethinking "Status" and "Role": Toward a General Model of the
        Cultural Organization of Social Relationships. In *The Relevance
        of Models for Social Anthropology,* M. Banton (ed.), pp. 1-24.
        New York: Praeger.

Gould, R. A.
1980   *Living Archaeology.* London: Cambridge University Press.

Gummerman, G. J. (ed.)
1971   The Distribution of Prehistoric Population Aggregates. *Prescott
        College Anthropological Reports* 1. Prescott.

Harris, M.
1971   *Culture, Man and Nature.* New York: Thomas Y. Crowell.

Hayden, B. (ed.)
   1979   *Lithic Use-Wear Analysis*. New York: Academic Press.

Higgs, E. S. (ed.)
   1972   *Papers in Economic Prehistory*. New York: Cambridge University
           Press.

   1974   *Paleoeconomy*. New York: Cambridge University Press.

Hill, J. N.
   1970   Broken K Pueblo: Prehistoric Social Organization in the Ameri-
           can Southwest. *Anthropological Papers of the University of
           Arizona* 18.   Tucson.

   1976   Individuals and Their Artifacts: An Experimental Study in
           Archaeology. *American Antiquity* 43(2):245-257.

Hill, J. N., and J. Gunn
   1977   *The Individual in Prehistory: Studies of Variability in Style
           in Prehistoric Technologies*. New York: Academic Press.

Hodder, C., and C. Orton
   1976   *Spatial Analysis in Archaeology*. London: Cambridge University
           Press.

Hole, F., and R. Heizer
   1973   *An Introduction to Prehistoric Archaeology*. New York: Holt
           Rinehart and Winston.

Hugh-Jones, S.
   1963   The Lord and the Regalia. *Horizon* 5(5):92-106.

Hunter-Anderson, L.
   1977   A Theoretical Approach to the Study of House Form. In *For Theory
           Building in Archaeology*, L. R. Binford (ed.), pp. 287-324. New
           York: Academic Press.

Huntington, R., and P. Metcalf
   1979   *Celebrations of Death: The Anthropology of Mortuary Ritual*.
           Cambridge: Cambridge University Press.

James, E. O.
   1928   Cremation and the Preservation of the Dead in North America.
           American Anthropologist 30:214-242.

King, L.
   1969   The Medea Creek Cemetery (LAn-243): An Investigation of Social
           Organization from Mortuary Practices. *Archaeological Survey
           Annual Report* 11:23-68. Los Angeles: University of California.

Klein, L. S.
   1977   A Panorama of Theoretical Archaeology. *Current Anthropology* 18:
           1-47.

Koerper, H.
   1976   *Systems Theory in Archaeology: Its History and Future*. Unpub-
           lished manuscript in possession of the author.

Kolata, G. B.
  1974   !Kung Hunter-Gatherers: Feminism, Diet and Birth Control. *Science*
          185(4155):932-934.

Kramer, C. (ed.)
  1979   *Ethnoarchaeology*. New York: Columbia University Press.

Kretchmer, N.
  1972   Lactose and Lactase. *Scientific American* 227(4):70-79.

Kroeber, A. L.
  1927   Disposal of the Dead. *American Anthropologist* 29:308-315.

Krupp, E. C. (ed.)
  1977   *In Search of Ancient Astronomies*. New York: Doubleday.

Lane, R. A., and A. J. Sublett
  1972   Osteology of Social Organization Residence Pattern. *American
          Antiquity* 37(2):186-201.

Leakey, M. D.
  1971   *Olduvai Gorge: Excavations in Beds I and II, 1960-63*. Vol. 3.
          Cambridge: Cambridge University Press.

LeBlanc, S.
  1973   Two Points of Logic Concerning Data, Hypotheses, General Laws,
          and Systems. In *Research and Theory in Current Archaeology*,
          C. L. Redman (ed.), pp. 199-214. New York: Wiley.

Lee, R. B.
  1972   Population Growth and the Beginnings of Sedentary Life Among the
          !Kung Bushmen. In *Population Growth: Anthropological Implica-
          tions*, B. Spooner (ed.). Cambridge: MIT Press.

Lee, R. B., and I. DeVore (eds.)
  1972   *Kalahari Hunter-Gatherers*. Cambridge: Harvard University Press.

Leone, M. P. (ed.)
  1972   *Contemporary Archaeology: A Guide to Theory and Contributions*.
          Carbondale: Southern Illinois University Press.

Lieberman, P., E. S. Creslin, and D. H. Klatt
  1972   Phonetic Ability and Related Anatomy of the Newborn and Adult
          Human, Neanderthal Man, and the Chimpanzee. *American Anthro-
          pologist* 74(3):287-307.

Linton, R.
  1937   One Hundred Per Cent American. *The American Mercury*, pp. 427-
          429.

Longacre, W. A.
  1970a   Archaeology as Anthropology: A Case Study. *Anthropological
          Papers of the University of Arizona* 17. Tucson.

Longacre, W. A. (ed.)
  1970b *Reconstructing Prehistoric Pueblo Societies*. Albuquerque:
         University of New Mexico Press.

Lotka, A.
  1925   *Elements of Physical Biology*. Baltimore.

Martin, P., and F. Plog
  1973   *The Archaeology of Arizona*. New York: Doubleday History Press.

McGee, W. J.
  1898   The Seri Indians. *United States Bureau of American Ethnology
         Annual Report, 1895-96*. Washington.

Meighan, C. W.
  1966   *Archaeology: An Introduction*. San Francisco: Chandler.

Miller, J. G.
  1965a Living Systems: Basic Concepts. *Behavior Science* 10:193- 237.

  1965b Living Systems: Cross-Level Hypotheses. *Behavior Science* 10:
         380-411.

Muller, J. W. (ed.)
  1975   *Sampling in Archaeology*. Tucson: University of Arizona Press.

Naroll, R.
  1962   Floor Area and Settlement Population. *American Antiquity* 27
         (4):587-589.

Oakley, K. P.
  1959   *Man the Tool-Maker*. Chicago: University of Chicago Press.

Odum, E. P.
  1963   *Ecology*. New York: Holt, Rinehart and Winston.

  1975   *Ecology, the Link Between the Natural and the Social Sciences*
         (2nd ed.). New York: Holt, Rinehart and Winston.

Onions, C. T. (ed.)
  1966   *Oxford Dictionary of English Etymology*. Oxford: University of
         Oxford Press.

Peebles, C. S.
  1977   Biocultural Adaptation in Prehistoric America: An Archeologist's
         Perspective. In *Biocultural Adaptation in Prehistoric America*,
         R. L. Blakely (ed.), pp. 115-130. *Southern Anthropological
         Society Proceedings* 11. Athens.

Pfeiffer, J. E.
  1977   *The Emergence of Society: A Prehistory of the Establishment*.
         New York: McGraw-Hill.

Plog, F. T.
  1975   Systems Theory in Archaeological Research. In *Annaul Review of
         Anthropology*, B. J. Siegel (ed.), pp. 207-224. Palo Alto:
         Annual Reviews.

Puleston, D. E.
    1971   An Experimental Approach to the Function of Classic Maya Chultuns.
            American Antiquity 36(3):322-325.

Raab, L. M.
    1973   Research Design for Investigations of Archaeological Resources
            in the Santa Rosa Wash: Phase 1. *Arizona State Museum Archaeo-
            logical Series* 26. Tucson.

    1977   The Santa Rosa Wash Project: Notes on Archaeological Research
            Design Under Contract. In *Conservation Archaeology: A Guide
            for Cultural Management Studies*, Michael Schiffer and George
            Gumerman (eds.). New York: Academic Press.

Read, D. W.
    1974   Some Comments on the Use of Mathematical Models in Anthropology.
            American Antiquity 39:3-15.

    1978   Appendix to Descriptive Statements Covering Laws, and Theories
            in Archaeology. *Current Anthropology* 19(2):307-335.

Read, D., and S. Le Blanc
    1978   Descriptive Statements Covering Laws, and Theories in Archaeo-
            logy. *Current Anthropology* 19(2):307-335.

    1979   Reply to More on Theory Building by E. G. Stickel. *Current
            Anthropology* 20(3):622-624.

Redman, C. L. (ed.)
    1973   *Research and Theory in Current Archaeology*. New York: Wiley.

Redman, C. L.; M. J. Berman; E. V. Curtin; W. T. Langhorne, Jr.; N. M.
            Versaggi; and J. C. Wanser
    1978   *Social Archeology: Beyond Subsistence and Dating*. New York:
            Academic Press.

Renfrew, C.
    1972   *The Emergence of Civilization: The Cyclades and the Aegean in
            the Third Millennium B.C.* London: Metheun.

    1975   *Beyond a Subsistence Economy: The Evolution of Social Organiza-
            tion in Prehistoric Europe*. Boston: MIT Press.

Renfrew, C., and K. Cooke
    1978   *Transformations: Mathematical Approaches to Cultural Change*.
            New York: Academic Press.

Renfrew, J. M.
    1973.   *Paleoethnobotany*. New York: Columbia University Press.

Revel, J. F.
    1970   *Ni Marx, Ni Jésus*. Paris: Robert Laffont.

Rodgers, W. B.
    1969   Environment and Change. *Anthropology UCLA* 1(1):1-13. Los
            Angeles.

Rouse, I.
  1972   *Introduction to Prehistory: A Systemic Approach*. New York:
          McGraw-Hill.

Rowlands, M.
  1978   Comments on Descriptive Statements Covering Laws, and Theories
          in Archaeology by D. Read and S. Le Blanc. *Current Anthropology*
          19(2):307-335.

Sabloff, J. A., and C. C. Lamberg-Karlovsky (eds.)
  1975   *Ancient Civilization and Trade*. Albuquerque: University of
          New Mexico Press.

Salmon, M. H.
  1978   What Can Systems Theory Do for Archaeology? *American Antiquity*
          43(2):174-183.

Saxe, A. A.
  1970   *Social Dimensions of Mortuary Practices*. Ph.D. dissertation,
          Department of Anthropology, University of Michigan.

Schalk, R. F.
  1977   The Structure of an Anadromous Fish Resource. In *For Theory
          Building in Archaeology*, L. R. Binford (ed.), pp. 207-249. New
          York: Academic Press.

Schiffer, M. B.
  1972   Cultural Laws and the Reconstruction of Past Lifeways. *The
          Kiva* 37(3):148-157.

Schiffer, M. B. (ed.)
  1978   *Advances in Archaeological Method and Theory 1*. New York:
          Academic Press.

Schneider, H. K.
  1977   Prehistoric Transpacific Contact and the Theory of Culture
          Change. *American Anthropologist* 79:9-25.

Semenov, S. A.
  1964   *Prehistoric Technology*. M. W. Thompson, translator. London:
          Cory, Adams and Mackay.

Service, E. R.
  1962   *Primitive Social Organization*. New York: Random House.

  1969   Models for the Methodology of Mouthtalk. *Southwestern Journal
          of Anthropology* 25(1):68-80.

Smith, M. A.
  1955   The Limitations of Inference in Archaeology. *Archaeological
          News Letter* 6:1-7.

South, S.
  1977   *Method and Theory in Historical Archaeology*. New York:
          Academic Press.

Spaulding, A. C.
  1973   Archaeology in the Active Voice: The New Anthropology. In
          Research and Theory in Current Archaeology, C. L. Redman (ed.),
          pp. 337–354. New York: Wiley.

Spoehr, A.
  1973   Zamboango and Sulu: An Archaeological Approach to Ethnic
          Diversity. *University of Pittsburgh, Department of Anthro-
          pology Ethnology Monographs* 1. Pittsburgh.

Spooner, B. (ed.)
  1972   *Population Growth: Anthropological Implications*. Cambridge:
          MIT Press.

Steward, J.
  1955   *Theory of Culture Change: The Methodology of Multilinear
          Evolution*. Urbana: University of Illinois Press.

Stickel, E. G.
  1968   Status Differentiation at the Rincon Site. *Archaeological
          Survey Annual Report* 10:209–261. Los Angeles: University of
          California.

  1979   More on Theory Building in Archaeology. *Current Anthropology*
          20(3):621–622, 624.

Stickel, E. G., and J. Chartkoff
  1973   On the Nature of Scientific Laws and Law-Building in Archaeology.
          In *The Explanation of Culture Change: Models in Prehistory*,
          C. Renfrew (ed.), pp. 663–667. London: Duckworth.

Tainter, J. A.
  1971   Salvage Excavations at the Fowler Site: Some Aspects of the
          Social Organization of the Northern Chumash. *San Luis Obispo
          County Archaeological Society Occasional Papers* 3. San Luis
          Obispo.

  1973   The Social Correlates of Mortuary Patterning at Kaloko, North
          Kona, Hawaii. *Archaeology and Physical Anthropology in Oceania*
          8:1–11.

  1975   *The Archaeological Study of Social Change: Woodland Systems
          in West-Central Illinois*. Ph.D. dissertation, Department of
          Anthropology, Northwestern University.

  1977   Modeling Change in Prehistoric Social Systems. In *For Theory
          Building in Archaeology*, L. R. Binford (ed.), pp. 327–351.
          New York: Academic Press.

  1978   Mortuary Practices and the Study of Prehistoric Social Systems.
          In *Advances in Archaeological Method and Theory*, M. B. Schiffer
          (ed.), pp. 105–141. New York: Academic Press.

Tainter, J. A., and R. H. Cordy
  1977   An Archaeological Analysis of Social Ranking and Residence
          Groups in Prehistoric Hawaii. *World Archaeology* 9:95–112.

Turner, G., and L. Lofgren
   1966   Household Size of Prehistoric Western Pueblo Indians. *South-
           western Journal of Anthropology* 22(2):117-132.

Ucko, P. J.
   1969   Ethnography and Archaeological Interpretation of Funerary
           Remains. *World Archaeology* 1:262-280.

von Bertalanffy, L.
   1950   An Outline of General Systems Theory. *British Journal of the
           Philosophy of Science* 1:134-165.

Watson, P. J., S. A. Le Blanc, and C. L. Redman
   1971   *Explanation in Archaeology: An Explicitly Scientific Approach.*
           New York: Columbia University Press.

Wedgwood, C. H.
   1927   Death and Social Status in Melanesia. *Journal of the Royal
           Anthropological Institute of Great Britain and Ireland* 57:377-
           397.

Weiss, G.
   1972   A Scientific Concept of Culture. *American Anthropologist* 75(4):
           1376-1413.

Weiss, K. M.
   1973   Demographic Models for Anthropology. *Society for American
           Archaeology Memoirs* 27.

Willey, G. R.
   1974   New World Prehistory. *American Journal of Archaeology* 78:321-
           331.

Willey, G. R., and J. A. Sabloff
   1974   *A History of American Archaeology.* San Francisco: W. H.
           Freeman.

Wilson, E. O.
   1975   *Sociobiology: The New Synthesis.* Cambridge: Belknap Press
           of Harvard University Press.

Wobst, H. M.
   1974   Boundary Conditions for Paleolithic Social Systems: A Simula-
           tion Approach. *American Antiquity* 39(2):147-178.

   1976a   Locational Relationships in Paleolithic Society. In *The Demo-
           graphic Evolution of Human Populations*, R. H. Ward and K. M.
           Weiss (eds.), pp. 49-58. London: Academic Press.

   1976b   Stylistic Behavior and Information Exchange. *University of
           Michigan Museum of Anthropology Anthropological Papers* 61:317-
           342. Ann Arbor.

Wright, G. A.
   1971   Origins of Food Production in Southwestern Asia: A Survey of
           Ideas. *Current Anthropology* 12:4-5.

Wright, S.
  1955   Classification of the Factors of Evolution. *Proceedings of the Cold Springs Harbor Symposia on Quantitative Biology* 20: 16–24. Cold Springs Harbor, New York.

Yellen, J. E.
  1977   *Archaeological Approaches to the Present: Models for Reconstructing the Past.* New York: Academic Press.